The Diamond
and
The Star

The Diamond
and
The Star

An exploration of
their symbolic meaning
in an insecure age

John Warden

SHEPHEARD-WALWYN (PUBLISHERS) LTD

First published in 2009 by
Shepheard-Walwyn (Publishers) Ltd
15 Alder Road
London SW14 8ER

British Library Cataloguing in Publication Data
A catalogue record of this book
is available from the British Library

ISBN: 978-0-85683-255-0

Typeset by Alacrity,
Sandford, Somerset
Printed and bound through
s|s|media limited, Wallington, Surrey

To Jung, the great rebel

'Socrates is charged with corrupting the youth
of the city, and with rejecting the gods of
Athens and introducing new divinities.'

Apologia, Plato

Contents

List of Plates

Foreword

IT GIVES ME great pleasure to introduce this inspiring and thoughtfully written book by Living Tao Foundation's long time Tai Ji friend, John Warden. I have known John for a number of years now and have been consistently amazed by his multi-dimensional interest in life and his ever-enthusiastic ways of pursuing his diversity of learning.

Here, in this gem of a book, he offers us the enrichment of many creative ideas from his lifelong exploration of the scientific, psychological, intellectual and spiritual realms.

I am honored to read about his personal understanding of the Taoist 'Mystery of the Gateway' and the *I Ching* wisdom inspired by reading my book, *Quantum Soup*. I am also delighted to see how easily he could run with my suggestions of playing with the Hexagrams as potent symbolic language, linking them with all the variations in the *I Ching*. By doing so, he has managed to delve deeper into the subconscious, and the 'collective unconscious' of all life's 'ever presence' of learning.

This book uses the multiple facets of the diamond to explore the metaphors of the solar system and reveal the underlying light of awakening consciousness. Congratulations, John!

CHUNGLIANG AL HUANG

Preface

JOHN WARDEN has written a fascinating book, at once a spiritual auto-biography and a psycho-cosmological adventure. Making deft use of symbolic resources from East and West, from Jungian Psychology and chemical analysis, the author constructs a taut but widely extended net of ideas which bring to our attention the need in the modern/post-modern world for a new kind of spirituality, transcendent yet inner-worldly, drawing on many ancient traditions, yet mindful of the awe-inspiring world that modern science has revealed to our gaze.

A key notion in all of this is Jung's idea of 'individuation'. Not every-one views this concept as having clearly spiritual content. The ideal of a well-balanced psychic life may seem to some as lacking in the kind of transcendence that they see as an essential ingredient of the spiritual life, perhaps too tied down to earth, to the chthonic, and to the collective unconscious. However, Jung's attempt to rethink the psychic life beyond Christian doctrine and beyond Freudian reductionism provides a power-ful stimulus to our imagination that has far wider significance than the typically psychological realm.

The burgeoning interest in a renewed but secular spirituality has much to gain from Jung's approach. The major feature of this new spiritual impulse lies in its eclecticism. Here Jung's approach shines forth strongly, in his concern to take heed of, if not actually to emulate, Eastern models of the spiritual life. It is evident too in his refusal to adopt the easy dichotomy between the scientific and the personal, and to find some way of integrating the new scientific paradigm which emerged in the twen-tieth century with a fuller account of the psyche. The psychic life could only be conceived as a whole, and as integrated in some way with the wholeness of the cosmos at large. And Jung's powerful contribution towards a fresh understanding of the important role of symbols in our search for meaning and integration is an important message for our times in which the public imagination has been dredged of many of its archetypal resources.

John Warden's own life path, which has evidently traversed many different fields of human experience and endeavour, has enabled him to approach in many creative ways the deep issues that concern our contemporary search for meaning and hope. His wide reading is very evident throughout this book, with its discussions ranging over issues of consciousness, Gaia theory, the *I Ching*, mandalas and synchronicity as well as many other ramifications of Jungian psychology.

But what shines through most to me is his ability to bring the world of science alive and to give it the kind of heart and soul, a deep symbolic significance, which to many it seems to lack. This is especially impressive in his penetrating discussion of the various transformations of what he calls 'the miracle of carbon ... the magic element on which life itself depends', an element which, like the Tao, is 'filled with infinite possibilities'.

This is an inspired book and I hope it will be widely read and meditated upon.

JOHN CLARKE
*Professor Emeritus in the History of Ideas
at Kingston University, London*

Author's Preface

SINCE THE INTRODUCTION which follows is rather long and discursive it was suggested that I preface the book with a brief explanation of its content and purpose. However I do not think this would be meaningful without a brief account of how it came to be written.

The book is an amplification of an essay written in 1988 at the end of a period of study and analysis at the Jung Institute near Zurich, a time of some emotional turmoil when I seemed in close touch with my unconscious and visited with visions and synchronistic events. Something – possibly contemplation of the Diamond Body, a Self symbol described in Richard Wilhelm's *The Secret of the Golden Flower*, made me wonder, as a chemist, why Jung, so concerned as he was with quaternity, and its expression in the Indian and Tibetan Self symbol of the mandala, had never seemingly recognised the important nature of the tetrahedron, both as the simplest form of solid, with four corners and faces, and its crystalline expression in the Chinese Self symbol of the Diamond Body. Again, as a chemist, I could visualise the huge symbolic importance of the quaternity represented by the four valences of carbon in the structure of all living matter.

This experience led me to draw what I have called the 'diamond mandala' and subsequently to follow it up with the 'star mandala' (Plate 1). It may interest some readers to start by reading the account of this experience which forms the second appendix. What prompted the drawing of the star mandala is difficult to understand, but having drawn it I recognised a close correspondence or association between the two symbols, diamond and star, which can be observed in so many emotional contexts, in poetry in particular. This seemed to me no coincidence, since they are such vital components of all life, related much as *yang* and *yin*, the star representing energy and the diamond matter. Life, particularly life on earth, in Chinese eyes the Tao, could be seen as the offspring of their marriage or symbiosis.

At the time in question my interest in Wilhelm's book (above) and his translation of the *I Ching* had led me into an exploration of Lao Tsu and to a workshop on Tai Ji where I met Chungliang Al Huang. I read Capra's *The Tao of Physics*; but it was after reading James Lovelock's *Gaia* and I could foresee the dangers to the planet, that the importance of the symbols for others than myself began to make itself felt. The original essay was only read by my analyst and a few friends. Now, some twenty years later, the crisis is manifest and is the topic of widest public concern. Much has been written and will continue to be written about our attitude to the planet which is our home, but we are dogged by the belief that the solution, if any, to the present crisis must lie somewhere in and among the various collectives – scientists and politicians and attendant lobbies – in which we, as individuals, can lay the responsibility and which we can comfortably blame when things go wrong. Jung may not have been the first to lay emphasis on the unconscious shadow in the individual psyche and its consequence in the phenomenon of the scapegoat, but if there was something which he held of the first importance it was the responsibility in every individual to understand himself. 'If the whole is to change,' he wrote, 'the individual must change himself.' Some fifty years after his death, Jung's ideas, the product of a lifetime of effort, have reached comparatively few. To understand Jung is impossible without experiencing for oneself the magic of the symbol. If this book has any central purpose it lies in the hope that readers may begin to recognise its place and purpose in their own lives. The diamond and the star are such wonderful symbols that I have tried to observe them in a variety of contexts, so that their magic may stimulate the imagination. Without this experience as part of the life of many more of us who, in the end, make up the responsible collectives, I fear for the future. In the end, the book is less about the diamond and the star than about the symbol itself. Many ideas will emerge, among them the close correspondence between the thinking of Jung and Lao Tsu and the emergence of a more image-oriented view of science, but the reader need not, perhaps should not, look for specific messages in each chapter so much as lean back and try to enjoy and integrate the kaleidoscope of images presented, and perhaps meditate on their personal meaning.   JOHN WARDEN 2007

Publishers' Note: The author sadly died before the book was published.

Introduction

*And furthermore, my son, be admonished: of making many books there is no end;
and much study is a weariness of the flesh.*   Ecclesiastes 12.12

CERTAINLY WHAT STUDY I have made has made me recognise in some despair how much has been written and what little hope there is to read even what I feel I should have taken into account. Yet the prophet also said, 'The words of the wise are as goads.' Who will ever read this book, I wonder? It is certainly vanity to suppose myself wise. And, however much the world needs them, it is not so much goads I seek to bring (others are trying) as more gently, I hope, to open some eyes.

This is not an academic book. Nor is it an autobiography or *apologia pro vita sua* – certainly not a sermon. Rather it is an attempt to show, or project, like an old fashioned lantern slide, some of the images and experiences and their background that have been instrumental in opening my own eyes, in fashioning what I am today.

Perhaps too much has been written and continues to be written in a rhetorical or evangelistic way, as though society can somehow be swayed into adopting some new or improved consciousness. The truth is that any such effort is doomed. As Jung so cogently put it:

If the whole is to change, the individual must change himself.[1]

In spite of his efforts to change society E.F. Schumacher also recognised it:

> It is no longer possible to believe that any political or economic reform, or scientific advance, or technological progress could solve the life-and-death problems of industrial society. They lie too deep, in the heart and soul of every one of us. It is there that the main work of reform has to be done – secretly, unobtrusively.[2]

How is an individual to change? Indeed, why ever should he? Only by some inspiration. So if this book has a purpose it can only be, however

1

grandiose a hope, to inspire; or rather to catalyse some inspiration within the reader.

When I first felt impelled to write about the diamond and the star, back in 1988, I was in a state of some turmoil at the end of a prolonged stay at the Jung Institute in Zurich and reconciling myself to the fact that continuing to pursue my ambition to become an analyst was a mistake. The decision would involve also an end to a long analysis with Dr Richard Pope for whom I had developed a great affection. Symbols came flooding in and I felt impelled to draw them as a first and then a second mandala (Plate 1). I then started to write in a kind of fever. As I wrote, a thesis of a sort took form and developed into a reconciliation of Eastern and Western approaches to psychology, Eastern being focused more on the Taoist tradition, and based largely on Jung's writings in Richard Wilhelm's translations of the *I Ching* (the ancient Chinese *Book of Changes*) and *The Secret of the Golden Flower*.

The original work was quite brief and made no concessions to a general readership. Some eighteen years later something impelled me to return to it. I feel keenly the new anxiety over the state of the planet and one of the seeds must have been a re-reading of James Lovelock's *Gaia: The Practical Science of Planetary Medicine*. Although my original intention is still evident, what now seems important is to use the insights granted to me then to illustrate and encourage a new attitude to life, whatever life and of whatever kind, remains to us and our descendants

There is a common thread extending through Jung's psychology and Taoism, and I believe it extends to the infant science of our planet and its system. In the three disparate disciplines lies an exhortation to an awakening of minds to a fresh understanding of ourselves, our habitat, and our place within it, and among its inhabitants, as whole entity. This cannot come about through a blinkered pursuit of reductive analysis, whether of psychology or the more objective sciences. The process of rational thinking through cause and effect has led to enormous strides in understanding, but at the cost of ever increasing division and sub-division to an extent that – to use one of the metaphors our language relies on so heavily – we can no longer 'see the wood for the trees'. Indeed so great are the divisions in science that it seems sometimes that we can barely discern trees among the myriad cellular species individually studied by different specialists. James Lovelock has put it more robustly:

Unfortunately, science is divided into a myriad of facets like the multi-lensed eye of a fly and through each separate lens peers a professor who thinks that his view alone is true.[3]

The divisions are not peculiar to science but extend through language and religion, the result of the remnants of tribalism prevailing in the human species.

Jung is probably best known for his advocacy of what he termed 'individuation' – the reintegration of suppressed contents of the unconscious psyche[4] to produce a whole or rounded individual. This is not an instant fix but rather a lifelong quest, never fully achievable, which can be assisted by an analyst although many highly individuated people, past and present, have never been near one. A first step is an acknowledgement and experience of the 'shadow' – the suppressed, unacknowledged inferior qualities and even evil within. I was interested to note that the only references to Jung in Mary Midgley's book *The Essential Mary Midgley* concern the shadow:

> The trouble is not, of course, that vanity is the worst of the vices. It is just that it is the one which makes admitting all the others unbearable, and so leads to the shadow-shedding project. And the reason why this project is doomed is because, as Jung sensibly points out, shadows have a function:
>
> *Painful though it is, this [unwelcome self-knowledge] is in itself a gain – for what is inferior or even worthless belongs to me as my shadow and gives me substance and mass. How can I be substantial if I fail to cast a shadow? I must have a dark side also if I am to be whole; and inasmuch as I become conscious of my shadow – I also remember that I am a human being like any other.*
> Quoted from *Modern Man in Search of a Soul*
>
> The acknowledged shadow may be terrible enough. But it is the unacknowledged one which is the real killer.[5]

What a great symbol is the shadow. It is the real killer of course because unconscious shadow contents are inevitably projected on to others as scapegoats. The true possessor never feels to blame. The shadow, constituting as it does the whole personal unconscious, is not wholly evil. It contains complexes (autonomous groupings) many of which have a collective nature – the various inherited and instilled prejudices of the family and the tribe. Seen in this light the well meaning efforts of institutions to 'eradicate' racial and religious prejudice are laughable.

While Jung continued his work on the neuroses of his individual clients he studied extensively and was arguably the greatest polymath of the twentieth century. He also wrote prolifically – his collected works extend to over twenty volumes. Many of his papers and lectures demonstrate his concern for the effect of neglect of the unconscious on the state of humanity which he regarded as the greatest of all dangers both to humankind and the planet. In 1944 he wrote:

> Psychology is the youngest of the sciences and is only at the beginning of its development. It is, however, the science we need most. Indeed, it is becoming ever more obvious that it is not famine, not earthquakes, not microbes, not cancer but man himself who is man's greatest danger to man, for the simple reason that there is no adequate protection against psychic epidemics, which are infinitely more devastating than the worst of natural catastrophes.[6]

This was of course written in the midst of World War 2 and reflects the focus of the period, but there is still no shortage of psychic epidemics. He felt that the only hope for the world lay in an increased consciousness, particularly of the shadow, in every individual. But he was far from optimistic. Towards the end of the same lecture he wrote:

> To be sure, a bloodless operation of this kind [bringing unconscious forces to consciousness] is successful only when a single individual is involved. If it is a whole family, the chances are ten to one against … But when it is a whole nation the artillery speaks the final word …
>
> *If the whole is to change, the individual must change himself.*

We might prefer these days to substitute sect or tribe for nation and bomb or machete for artillery, yet his words ring true. And they ring true not only for epidemics of a purely psychic nature. The following passage may be prescient:

> When mankind passed from an animated Nature to an exanimated Nature, it did so in the most discourteous way: animism was held up to ridicule, and reviled as a superstition. When Christianity drove away the old gods, it replaced them with one God. But when science de-psychised Nature, it gave her no other soul, merely subordinating her to human reason …

What science has discovered can never be undone. The advance of truth cannot and should not be held up. But the same urge for truth that gave birth to science should realise what progress implies. Science must recognise the as yet incalculable catastrophe which its advances have brought with them.[7]

Compare Schumacher, writing from a Christian point of view in 1979, while the Soviet Union was still a threat:

It may or may not be right to 'ban the bomb'. It is more important to overcome the roots out of which the bomb has grown. I think these roots are a violent attitude to God's handiwork instead of a reverent one. The unsurpassable ugliness of industrial society – the mother of the bomb – is a sure sign of its violence.[8]

I was interested to see Mary Midgley's comment in *Science and Poetry*, 'Gods are much easier to remove than demons.'[9] How true. Bringing to consciousness unpleasant personal episodes and prejudices is not merely an exercise of the mind or even the will; it must involve their re-experience, and suffering the feelings involved, whether of shame, anger or sadness. It will involve the withdrawal of long held projections, and this is experienced as what one might call 'withdrawal symptoms' so obvious in teenagers. It will affect not only the protagonist but all those involved with his or her projections. But an awakening consciousness brings its rewards in the form of a fresh outlook on life and fresh psychic energy.

In Jung's psychology unconscious contents are brought to the light of consciousness through symbols – the language of the unconscious, which the rational function of the mind must interpret. Symbols are made apparent through psychic experiences, particularly dreams, but also waking fantasies and strange coincidences, called by Jung 'synchronistic' events, and always having some symbolic character. It is through reflection on the symbol that the true recognition of the event and its significance comes about.

I hope to show in this volume the value of symbols, and their importance in our lives. They are fundamental to Jung's contribution to human wisdom, directed as it is to Western thought. Jung's first major work is entitled in English *Symbols of Transformation*,[10] and his collections of essays and lectures *The Symbolic Life*.[11] A symbolic life may be a step forward the planet needs.

While my own experience has been centred around Jung's contribution to psychology, I do not mean to undervalue that of Freudian psychoanalysis, especially in its modern movement. I have found a great deal of inspiration and support in Margaret Arden's *Midwifery of the Soul*,[12] which has the subtitle 'A Holistic Perspective on Psychoanalysis', to which I shall be referring. Many of my references to Jung might equally refer to 'depth psychology'.

It is through symbols that the thread reaches out to Eastern thought, and to planetary science which I believe is showing the way to a more comprehensive and less hubristic approach to the quest for scientific truth. Planetary science, perhaps now, to mimic Jung's words, the youngest of the sciences, embraces and endeavours to integrate all the principle divisions: chemistry, physics, biology and geology. It cannot afford to ignore psychology. Yet it must beware, here as elsewhere, of being lured down specialist pathways, many of which have lost sight of their origins.

In the West the Abrahamic religions pursued the conscious/unconscious split, emphasising its attendant polarities, in particular sin and virtue, affirming a quest for perfection at the expense of the shadow and dividing inevitably into numerous sects each propounding its own belief as the only truth. It was left to the underground, the Kabbala and in particular the alchemists, to seek out bravely an alternative path – the *Magnum Opus* of transmuting the base metal to gold. Since the alchemists pursued their 'science' in fear of their lives, they made full use of symbols: the *prima materia* – the original state, often symbolised by a joined male/female figure signifying the opposites, tramping on putrefying remains or excrement (Plate 3a). and the *hieros gamos* or *conjunctio* – the final sacred marriage, signifying the integration of the opposites and leading to the Philosopher's Stone and finally the Elixir of Life, the end of their quest (Plate 4a). As Jung realised and described in his books *Psychology and Alchemy* and *Mysterium Conjunctionis*, this was in truth a psychological quest for individuation, although hidden in magical codes and rituals and confused by charlatans. Tramping the putrefaction was experiencing the shadow.

Eastern history is very different; so different in fact that the West has difficulty in empathising with its thought and outlook. The split to consciousness has never been sharp and even now it seems to proceed from a limbo region.

Western thought and attitudes have been subjected to a continuous influence from the East since Marco Polo in the middle ages, although the first significant interaction was with the Jesuit missionaries in the seventeenth century. However the reception of this influence has fluctuated very considerably. Goethe, and later the German philosophers, accepted it avidly, as did much of Victorian England, but with the advent of increasingly linear science and linguistic philosophy it fell into disrepute. Recently there has been a resurgence in interest. John Clarke[*] has provided an extensive survey of the history of Oriental influence in his trilogy of books, *Oriental Enlightenment, Jung and Eastern Thought* and *The Tao of the West*. At the end of the historical survey, *Oriental Enlightenment* (1997), he concludes:

> What we are witnessing today is a pandemic transformation of ideas and institutions, led by a cultural and political energy which had its origins in the West, but which now extends world-wide in its scope and influence. The long historical process of planetary fusion ... is even today growing apace in all kinds of fields of intellectual, cultural, and political endeavour. Future historians may view this as marking the end of the ancient division of East and West, and the end of orientalism. On the other hand they may see it as the beginning of a new phase of orientalism – or whatever it might be ...

In his second book, Clarke discusses the influence of Oriental thought on the work of C.G. Jung:

> Its romantic impatience with rational and conventional structures and its eagerness to make contact with the primordial sources of life was immediately attractive to him. He was drawn, too, to its dynamic cosmology based on the idea of ever-changing and ever-interacting forces, characterised in terms of energy (ch'i) which flows through the whole of nature, human and non-human. In the central concept of tao (the way) he saw a close affinity with his idea of synchronicity, and in the idea of wu-wei (action through non-action) he recognised a psychological attitude which ran closely parallel to his own approach to the unconscious mind. And finally the concepts of yang and yin, opposing yet complementary principles that underpin all of reality and human experience, matched with remarkable exactness his conception of the psyche as a self-balancing system governed by the tension of opposing principles.

[*] Professor Emeritus in the History of Ideas, Kingston University, Surrey.

Towards the end of his last book (2000) he points to factors representing ways of thinking affecting the popularity of Taoism in the West. The first is 'the desire to discover an alternative or transformed spirituality or religiosity without credal commitment or doctrinal validation, a sense of meaning without transcendence or teleology, a non-theistic spirituality which is contained within our finite existence, is concerned with how to make the best of our life in this world and gives us a secure and life-giving anchorage in the natural world.' As aspirations these may be true, but I believe it would be a mistake to seek in Taoism anything resembling 'a secure anchorage', as Alan Watts has explained so well in his book *The Wisdom of Insecurity*.

The second factor to which Clarke points is 'a need which revolves around a holistic attitude of mind and which seeks fulfilment through the overcoming of the dualism of body and spirit.'

'Finally,' he continues, 'it is a need (within intellectual currents) to retreat from the certainties sought by positivists and a compensating emphasis in science, philosophy and literature on unpredictability, disorder, incommensurability and a suspicion of the truth-telling powers of language.'

Clarke concludes:

> At a more personal level, Daoism may become in the years ahead what Buddhism has already become, a serious option for the spiritually disenchanted and spiritually seeking, or at least an inspiration towards the emergence of a new pluralistic syncretism, a new form of spirituality which draws together elements of Daoism and other Asian religions to bring them, not only into dialogue, but also into active symbiosis with indigenous Western traditions and thinkers. This symbiosis is likely to have a strong affinity with green thinking, with the concern not just for a personal way of salvation but for the future of the planet, a counter to excessive consumption, materialism, environmental degradation and, in a word, a new way of thinking about our relationship with the natural world. This implies a non-exploitative relationship with the earth and with non-human creatures, and the development of technologies which go with rather than against nature ...

Jung recognised that for the ancient Chinese the unconscious was the natural condition, consciousness being seen as an archetypal affect – a kind of neurosis. Symbolic expressions were the natural order, at least in China where the language is written in images or ideograms. The old proverb: 'A picture is worth a thousand words' illustrates the Chinese standpoint. Lao Tsu in his *Tao Te Ching* moves gracefully from one image to the next:

> The fundamental Chinese idea of the order of nature is not compatible with formulation in the order of words, because it is organic, and is not linear pattern. Alan Watts[13]
>
> The tao that can be told
> Is not the eternal Tao[14]

Six hundred years later Jesus had to rely on parables. Taoism is therefore hard to express, though Alan Watts performed miracles in this respect, as did Richard Wilhelm earlier. Lao Tsu was probably the first to express in words the idea of a conscious/unconscious split, creating the pairs of opposites, although this derived from the earlier ideas compiled in the *I Ching*, the oracular *Book of Changes*. Yet Taoism is individuation expressed in its own language, contemplating for the adept a life lived in a state of conscious harmony with the natural order; following the grain of the wood; flowing with the current; never provoking fate by forced acts of will.

In Hinduism and Buddhism also, the unconscious is held in reverence. Their great symbol was the mandala, the 'squared circle', the circle signifying the 'all' and the square the material pattern of life, with the opposites indicated at the sides or corners pointing to the Godhead at the centre (Plate 2). (Rather strangely the alchemists also adopted this symbol.) These would be contemplated at length by adepts and postulants seeking an enlightened state of a harmonious relation with the unconscious. Sometimes mandalas were constructed as buildings called stupas around which pilgrims would walk. Buddhism later incorporated, in the Zen tradition, much derived from Taoism.

James Lovelock was inspired by the wonderful views of the planet earth experienced by the astronauts to perceive her as a living entity, and humanity not merely as inhabitants but one of a host of co-operating subjects: the animals and the trees, the seas and the clouds. From his friend William Golding he drew for her the name Gaia – the Greek goddess Earth. And from this intuition he has, with the help of his adherents, developed his ideas into a new science, effectively a holistic[15] physiology. One can no longer call this 'a branch of science', rather its root.

I notice that Lovelock, in his writing, seems to avoid the use of the term 'symbol', preferring 'metaphor'. Possibly this linguistic expression is more acceptable, used as it is more widely in philosophy, but the vision of Gaia was supremely a symbol, certainly as used by Jung as denoting the best possible expression of something new and unknown, and pregnant with meaning. Now Gaia has given birth. Lovelock is seeking a way of interpreting scientific data in a holistic way, gathering the expertise of the specialities – necessary as they are – rather like (I hope Lovelock will forgive me!) a mother hen gathering her chicks. But how is one to comprehend the clucking of the hen? For this a new language seems to be needed. It would be an interesting exercise to write everything in Chinese, but I imagine even to the modern Chinese, while flattering, it could not be realistic.

At the very least a new outlook is needed which gives each speciality its full due, but retains with a new humility the broad – so very broad – context. It is not of course possible, in the foreseeable future, to express science in anything other than its well-known conventions; but to hold the 'myriad things'[16] in mind requires, I believe, what I would call a symbolic attitude. If there is a unifying purpose in this book it is to explain and expound this idea. What I learned back in 1988 from drawing the two mandalas was the importance of *seeing in three dimensions*. This requires a mind opened and tuned to intuition, an imagination ready to explore the intuitions and a thoughtful intelligence willing to give them meaning.

Great scientists understand and yield homage to the power of intuition:

> Thought is only a flash between two long nights, but this flash is everything.
>
> Henri Poincaré

Kekule was inspired to discover the structure of benzene through a vision which (according to some reports) came to him while stepping off a bus.

But very often we need a midwife between the images of the unconscious mind and the linear language of conscious thought. The arts world is replete with these great interpreters, but who better than those wizards of words, the poets. In the chapter 'The Diamond and Star in Poetry' I have drawn from some of my own favourites, particularly to illustrate the polarities which are so much a feature of conscious existence.

I would like to return to Lovelock's picture of the fly with its multi-lensed eye. Each of these lenses brings in a different image. But there is only one fly – and he sees them all! He even has two such eyes. Maybe we need to evolve to a flynian state with advanced stereoscopic vision. But there is no time now for this evolutionary step. We must do the best we can as what we are and with what we have. And what we are is indeed a marvel. A live human being. An animal entity with mysterious inter-connections that allow an almost completely unconscious progression through a period of time while blessed with an ability to interpret, in a limited but sufficient way, its own environment. And our environment, what we have, is Gaia.

To go back to the beginning, I first saw in the diamond and the star a combined symbol which was the inspiration for the creation of a new form, 'three-dimensional' mandala marrying Eastern and Western psychology. I can now see in them another symbol of Gaia, the two-fold system of Earth and Sun. For it is a *binary* system. The Earth is only Gaia when illuminated, energised, by the Sun. The symbol, rejuvenated, is the inspiration for the present creative effort.

NOTES

1 C.G. Jung, *Collected Works* 18, 1358 [Epilogue to *Modern Man in Search of a Soul*].
2 E.F. Schumacher, *Good Work*, Abacus, 36.
3 James Lovelock, *Homage to Gaia*, Oxford, 3.
4 The use of the words 'psyche' and 'psychic' should not be given any 'extrasensory' connotation. They are used in psychology to distinguish matters of the mind from those of the body ('soma').
5 Mary Midgley, *The Essential Mary Midgley*, Routledge, 107.
6 C.G. Jung, *ibid.*
7 *Ibid.*, 8, 1368, 1377.
8 E.F. Schumacher, *Good Work*, Abacus.
9 Mary Midgley, *Science and Poetry*, Routledge, 33.
10 C.G. Jung CW 5.
11 *Ibid.*, CW 18.

12 Margaret Arden, *Midwifery of the Soul,* Free Association Books.
13 Alan Watts, *What is Tao,* New World, 85.
14 The opening of the *Tao Te Ching.*
15 The word 'holistic' has been open to much criticism. Stephen Rose in *Lifelines* prefers 'top-down' as opposed to 'bottom-up' for reductionist, but it seems apt to me.
16 The opening of the *Tao Te Ching.*

1

The Quality of Mystery

What we peer at and cannot see we call dim;
What we listen for and cannot hear we call faint;
What we grope for and cannot feel we call intangible;
These three are so imperceptible,
They cannot be distinguished, but blend into one.
In the light it does not sparkle;
In the dark it is not gloom.
Its boundaries cannot be defined,
But reach out to non-existence.
It has the shape of the shapeless,
And the image of the unimaginable,
Mysterious.
Stand before it and you will not see its face.
Follow it and you will not see its back.
Yet by keeping to the Tao of the past,
You will begin to master the present;
For as the present is to the past,
So is the past to the beginning of all things.
This is called following the thread of the Tao.

So SAYS Lao Tsu in the fourteenth chapter of the *Tao Te Ching,* The words are my own adapted from various translations. I make no apology. As Chungliang Al-Huang points out in his *Embrace Tiger, Return to Mountain,* there are some thirty English translations:

When I try to translate *Tao Te Ching* for you, I go through a helplessness of grabbing onto all the words I can find and then looking at your eyes and hoping that when I say 'female', something clicks for you ...[1]

For all their difficulty in translation the words of Lao Tsu vividly expound the eternal quality of mystery. Mystery will be with us always.

Whenever we approach an understanding of some aspect of life, there it remains, up ahead. If it should disappear, what would remain? A sterile world, I think, where all creation has disappeared.

The mysterious and the sense of wonder which it inspires serve the imagination as a trigger, and it is the imagination which is the source of all creative energy. Life, and the place of mankind within it, is itself the greatest mystery, impelling a need to worship, the source of all religion:

> When I consider the heavens, the work of thy fingers,
> The moon and stars, which thou hast ordained;
> What is man, that thou art mindful of him?
> And the son of man, that thou visitest him,
> For thou hast made him but little lower than God.[2]

Various authors including David Tracey[3] have pointed out, if there is any need, that university students, the open face of the present age, predominantly deny any interest in dogmatic religion; yet nearly all acknowledge a sense of the sacred or numinous, of the presence of a spiritual dimension. The world, it seems, has shaken off a phase of denial of the spiritual, of adherence to a practical humanism and has experienced an awakening. It was not so long ago, a mere matter of the sixties that I remember Bertrand Russell as the great student guru.

It may be an accompaniment to a reaction from the gradually appearing realisation of the despoliation of the planet and its resources, now reaching a crescendo. Yet from even earlier, I think, there has been a withdrawal of projections of omniscience from science and in particular medicine.

My school infirmary had a stock of ancient schoolboy literature and I remember as a child reading an old *Boy's Own Annual* of the 1920s giving the writer's impression of life in the next century. Among the sketches of men and women in short tunics I remember particularly a fantasy conversation between a boy and his father in which the boy was asking what was a 'cold'. The father explained that in the bad old days the world was subject to all kinds of illnesses of the body, but that with the march of science these were now a thing of the past.

We no longer hold to a belief that science will provide eventually the key to a disease-free world. No longer are we conditioned, when we fall ill, to place ourselves with little or no hesitation in the hands of a GP in the expectation of a cure. The final disillusionment was probably the advent of Aids.

Like all withdrawal symptoms they tend to impose a Heraclitan reaction towards an opposite pole. These heralded the advent of the New Age, with its focus on alternative and holistic medicine and organic food. This has been accompanied by a renewed interest in things pertaining to the spirit, originally perhaps in the occult but now seemingly more generalised.

The present struggle of the Christian establishments has been put by David Tracey down to a loss of vitality in the symbols which were once meaningful but have been devalued by the modern focus on 'ideas' and 'explanations', with consequent loss of mystery in religious services. I have long thought this true, but not wholly. The older tradition, where the religious service was to provide an *ambience*, where spiritual communication can thrive, whether with an objective God or an inner self, still holds true in places, and at times – especially at Christmas. The mystery of the incarnation remains a mystery still, however much effort is made to explain it away, and the older hymns and psalms, where permitted, can still weave their magic. The magic evolves from a re-ignition of the old symbology, experienced with all the senses, though particularly the ears, in the poetry and the music. At the centre or focus of this mystery is another mystery, the mystery of love, which all can recognise but no one fully explain.

Of all the mysteries that science has opened to us perhaps the one inspiring the greatest awe and that embodies so many of our ancient myths is the great universe or cosmos in which we find our insignificant place. In a remarkable book, *The View from the Centre of the Universe*, Joel Primack and Nancy Ellen Abrams have developed a human-centred view almost reversing the Copernican revolution:

> Most of us have grown up thinking that there is no basis for our feeling central or even important to the cosmos. But with the new evidence it turns out that this perspective is nothing but a prejudice. There is no geographic centre to an expanding universe, but we are central in several unexpected ways that derive directly from physics and cosmology – for example we are in the centre of all possible sizes in the universe, we are made of the rarest material, and we are living at the midpoint of time for both the universe and the earth.[4]

They point out that any picture of the universe can only be symbolic. 'No one can step outside of it to look at it.' The importance of symbols is possibly best recognised by astronomers, or, in the new language, cosmologists, together with the nuclear physicists, used as they are to handling extremes of dimensions far outside normal experience.

> Symbols are the way to grasp the universe. They free us from the limits of our five senses, which evolved to work in our earthly environment. Symbols let us see the universal with our mind, which doesn't have any obvious limits. And symbols are far easier to remember than a long, logical argument or a mathematical equation. Carl Jung wrote, 'A symbolic work is a perpetual challenge to our thoughts and feelings because even if we know what the symbols are, they do not refer to a given thing, like a sign, but are "bridges thrown out towards an unseen shore".'[5]

To represent their inside picture of the cosmos, they return to one of the earliest symbols known – the Uroboros, described in the introduction, and redraw it to show their new values (Plate 4b). This opens up a sense of the miraculous. It transpires that life occurs within a small zone at the midpoint of the whole time-span of the cosmos – from its beginning to its end – as now understood. The authors call it *Midgard* from the Norse. Furthermore it could only have occurred at this point, one of the constraints being the speed of communication required for consciousness. But *Midgard* does not only represent a midpoint in time, but the midpoint of a finite scale of orders of magnitude which the cosmos requires, from the scale of the smallest particle to the largest galactic supercluster. Again, life requires for its existence the rarest components of the cosmos, the complex elements they call stardust.

It might seem that since so much is known, there is no more room for mystery. Science seems to be approaching that extraordinary goal, 'a theory of everything'. The authors, indeed, seem sometimes to be bordering on hubris:

> The age of the universe is the linchpin of our understanding ... In 1998 the discovery of convincing evidence for dark energy ended the uncertainty. Dark energy exists. All discrepancies have disappeared. The universe is about 14 billion years old. We finally understand how the cosmos fits together.[6]

Yet the miracles inherent in the new Uroboric cosmology maintain or even expand the quality of mystery.

Both the ancient world and the Kabbalists were right about what lies beyond the known universe: we are surrounded by something mysterious and other. But unlike the ancient watery chaos, eternal inflation is far from inert. *So awesome is the energy of eternal inflation that the smallest possible amount of it, so small that it fell below eternity's minimum threshold and disappeared through the floor – that subminimal bit became our entire universe.* And the mysterious no longer surrounds just the earth and its little patch of visible heavens: eternal inflation surrounds the entire universe that was created by the Big Bang. Science has pushed back the borders of mystery and expanded the known universe immeasurably.[7]

To experience a symbol is to acknowledge and validate a mystery. For Primack and Abrams the New Uroboros was their inspirational symbol in whose mystery they sought their meaning, just as for me it lies in the symbiosis of the diamond and the star. But symbiosis is a cold word. In truth the relationship between the diamond and the star is an age-long love affair, well expressed in the Christmas hymn:

> *And man at war with man hears not*
> *The love song which they bring:*
> *O hush the noise, ye men of strife,*
> *And hear the angels sing.*[8]

But above everything, the great mystery of *purpose* remains: the impossible paradox 'why am I *me*?' of self-consciousness, which will be developed in another chapter.

Mystery, its validation and experience in symbols, is the central theme of this book.

NOTES

1 Chungliang Al-Huang, *Embrace Tiger, Return to Mountain,* Celestial Arts, 163.
2 Psalm 8, Revised version.
3 David Tracey, Lecture to the Scientific and Medical Network, 2006.
4 Joel Primack and Nancy Ellen Abrams, *The View from the Centre of the Universe,* Fourth. Estate, 7.
5 *Ibid.,* 11.
6 *Ibid.,* 132.
7 *Ibid.,* 204.
8 'It Came Upon the Midnight Clear', Christmas hymn.

2

The Mystery of Being

*Not how the world is, but that it is, is the mystical.**

THE WORD *mystery* combines, perhaps, the banal conception of paradox with that of the numinous, which requires a sense of wonder. The most striking example, to my mind, is the first, childhood, experience of self-consciousness, what Gabriel Marcel called 'The Ontological Mystery'. This can affect a lifetime's thought and attitude. For some it is closely bound to the mystery of God. For many also it is closely bound to the mysteries of morality and love. This chapter looks at four great thinkers, Gabriel Marcel, Jung, Iris Murdoch and (briefly) Teillard de Chardin, to see, and to an extent speculate, how the ontological mystery affected their lives and work.

The childhood experience can be so awe-inspiring as to approach the traumatic. This happened to me as a boy of seven. I was at a boarding school in the highlands of Malaya and absorbed in some primitive gardening in a jungle clearing – quite a mysterious ambience. I was totally self-absorbed, looking at my hands digging and crumbling the earth, finding myself thinking: 'What are these things, they look like hands, but they are my hands? They are part of me. But why? Why am I me? Why am I kneeling here, in this clearing, on this earth?'

The mystery was suddenly too great to handle. I felt my mind spinning. I was somewhere above, looking down at myself. The experience was frightening and I 're-entered myself' with a jolt.

This experience of self-consciousness has been nobly faced by Gabriel Marcel in his essay entitled 'The Ontological Mystery'. I do not know if

* Quoted by Iris Murdoch.

Marcel had a psychic experience akin to my own. Near the end of an autobiographical essay he writes:

> This leads me to believe that the development of my thought was largely an explication. It all seems to me to have happened as though I had only gradually succeeded in treating as material for thought what had been an immediate experience.[1]

I could say the same thing!

He writes in philosophic terms about the general concept of 'being', but from the eloquence of his language I should be surprised if he had not. He starts with a negative view:

> Rather than to begin with abstract definitions and dialectical arguments which may be discouraging at the outset, I should like to start with a sort of global and intuitive characterisation of the man in whom the sense of the onto-logical – the sense of being – is lacking, or, to speak more correctly, of the man who has lost the awareness of this sense ... The characteristic feature of our age seems to me to be what might be called the misplacement of the idea of function, taking function in its current sense which includes both the vital and the social functions. The individual tends to appear both to himself and to others as an agglomeration of functions.

He then goes on to show the need for this sense of being and to deny the likelihood of ever reaching a rational explanation:

> In such a world the ontological need, the need of being, is exhausted in exact proportion to the breaking up of personality on the one hand and, on the other, to the triumph of the category of the 'purely natural' and the consequent atrophy of the faculty of *wonder* ...
>
> Such a need, it may be noted, is to be found at the heart of the most inveterate pessimism. Pessimism has no meaning unless it signifies: it would surely be well if there were being, but there is no being, and I, who observe this fact, am therefore nothing.

Then we come to what is, in my mind, the root problem – self-consciousness:

> These preliminary reflections on the ontological need are sufficient to bring out its indeterminate character and to reveal a fundamental paradox. To formulate this need is to raise a host of questions: Is there such a thing as being? What is it? etc. Yet immediately an abyss opens under my feet: I who

ask these questions about being, how can I be sure that I exist? Yet surely I, who formulate this *problem*, should be able to remain *outside* it – *before* or *beyond* it? Clearly this is not so ... So I am inevitably forced to ask: Who am I – I who question being? How am I qualified to begin this investigation? If I do not exist, how can I succeed in it? And if I do exist, how can I be sure of this fact? Contrary to the opinion which suggests itself at this point, I believe that on this plane the *cogito* cannot help us at all.

Marcel speaks, of course, as a Christian and a Catholic, yet in his summing up he is at pains to declare that recognition of the ontological mystery is not confined to religion. Nevertheless it is clear that he feels that in some mysterious way it 'points towards the light':

> I would say that the recognition of the ontological mystery, in which I perceive as it were the central redoubt of metaphysics, is, no doubt, only possible through a sort of radiation which proceeds from revelation itself and which is perfectly well able to affect souls who are strangers to all positive religion of whatever kind; that this recognition, which takes place through certain higher modes of human experience, in no way involves the adherence to any given religion; but it enables those who have attained to it to perceive the possibility of a revelation in a way which is not open to those who have never ventured beyond the frontiers of the realm of the problematical and who have therefore never reached the point from which the mystery of being can be seen and recognised. Thus, a philosophy of this sort is carried by an irresistible movement towards the light which it perceives from afar and of which it suffers the secret attraction.

Marcel's own psychic experiences ended with a Christian affirmation. To this extent I cannot quite keep pace. My own experience left me simply with a sense of a huge and fundamental mystery. However the experience, if I am in any sense typical, is one that is never forgotten and moreover frequently recurs at odd moments, usually when least expected, although in a muted form.

As to his 'movement towards the light', this seems to me something different. It is the old moral dilemma: is there a will towards goodness? If so how and where does it originate? If it does not exist, is there any hope for humankind, or indeed the planet? We are almost compelled to affirm its existence. But its origins, like being, are inexplicable from rational thought. It is another mystery.

As a disciple of Jung I have found this a difficult area in his writings. At some points he seems to affirm a moral position; at others he seems quite detached. When John Freeman asked him in a recorded interview, 'Do you believe in God?', he answered, 'I do not believe – I *know*.' He left this in the air with a humorous smile. He was in one sense pointing to his Gnostic roots – he would not tie himself to any dogmatic belief structure: he knew something, or he did not.

However there is more to it. A great deal of insight can be gained from his autobiographical work *Memories, Dreams, Reflections*. Jung's childhood experience of the ontological mystery was closely associated with his difficulties with the concept of God. He had an experience which lasted all his life and left him in no doubt of the existence of God – his own God, and rather different from the general conception of an all-loving being. He was rather more the all-terrible. For some time Jung had felt an experience approaching which he kept resisting. He felt it to be something against his upbringing, sinful. Finally he could hold out no longer:

> But from the moment I emerged from the mist and became conscious of myself, the unity, the greatness, and the superhuman majesty of God began to haunt my imagination ...
>
> Hence there was no question in my mind but that God Himself was arranging a decisive test for me ...
>
> I gathered all my courage, as though I were about to leap forthwith into hell-fire, and let the thought come. I saw before me the cathedral, the blue sky. God sits on His golden throne, high above the world – and from under the throne an enormous turd falls upon the sparkling new roof, shatters it, and breaks the walls of the cathedral asunder.
>
> So that was it! I felt an enormous, an indescribable relief. Instead of the expected damnation, grace had come upon me, and with it an unutterable bliss such as I had never known. I wept for happiness and gratitude. The wisdom and goodness of God had been revealed to me ...
>
> That was what my father had not understood, I thought ... he had failed to experience the will of God ... He had taken the Bible's commandments as his guide; he believed in God as the Bible prescribed and as his forefathers had taught him. But he did not know the immediate living God who stands, omnipotent and free, above His Bible and His Church, who calls upon man to partake of His freedom, and can force him to renounce his own views and convictions in order to fulfil without reserve the command of God.[2]

There is little doubt that, from that moment on, God was something that required no belief; that in a sense denounced belief; that was nothing if not an immediate experience. This was the God of Job, as brought out later in his *Answer to Job*:

> Behold, I am of small account; what shall I answer thee?
> I lay my hand on my mouth.[3]

'And indeed, in the immediate presence of the infinite power of creation, this is the only possible answer for a witness who is still trembling in every limb ...'[4]

It does appear that the experience was part and parcel of an experience of self-consciousness. As he developed his ideas Jung's conception of God was bound up in his writings on the Self, the origin of the archetypes, at the very root of the personal and collective unconscious. As borne out in his commentary on *The Secret of the Golden Flower*, it is not very different from the Tao. God was the All but with a more demanding presence than is evident in the *Tao Te Ching*.

What remains a puzzle is his (and indeed Lao Tsu's) accommodation of good and evil. There is very little room for this in a conception of balanced polarities. One Taoist philosopher wrote: 'The Tao is easy if one has no preferences.' Ambiguous indeed!

Later in the same passage Jung says:

> In His trial of human courage God refuses to abide by traditions, no matter how sacred. In His omnipotence He will see to it that nothing really evil comes of such tests of courage. If one fulfils the will of God one can be sure of going the right way.

This may be very well for a childhood experience, but it does not square with his life's work. The answer lies in *Answer to Job*, the writing of which amounted to his own wrestle with the angel. In the book Jung did not hold back from writing with full emotion, so much so that he did not authorise the book's publication until many years later when his reputation was more secure.

> Since I shall be dealing with numinous factors, my feeling is challenged as much as my intellect. I cannot, therefore, write in a coolly objective manner, but must allow my subjectivity to speak.[5]

Jung saw in the Book of Job the revelation of a God who was not only a unity of polarities, and therefore unconscious, but a God emerging into consciousness; emerging through the development of human consciousness. And he saw in this a meaning behind the Christian myth of the incarnation.

> Although the birth of Christ is an event that occurred but once in history, it has always existed in eternity ... What exists in the pleroma as an eternal process appears in time as an aperiodic sequence, that is to say it is repeated many times in an irregular pattern ...
>
> It was only very lately that we realised (or rather, are beginning to realise) the God is Reality itself ...[6]

Through man came the recognition of the polarities and with it the imperative of choice, the moral imperative. Possibly the best discussion I have found is in the penultimate chapter of *Memories, Dreams, Reflections*, 'Late Thoughts'.

> In any case, we stand in need of a reorientation, a *metanoia*. Touching evil brings with it the grave peril of succumbing to it. We must, therefore, no longer succumb to anything at all not even to good. A so-called good to which we succumb loses its ethical character. Not that there is anything bad in it on that score, but to have succumbed to it may breed trouble. Every form of addiction is bad, no matter whether the narcotic be alcohol or morphine or idealism. We must beware of thinking of good and evil as absolute opposites. The criterion of ethical action can no longer consist in the simple view that good has the force of a categorical imperative, while so-called evil can resolutely be shunned. Recognition of the reality of evil necessarily relativises the good, and the evil likewise, converting both into halves of a paradoxical whole.
>
> In practical terms, this means that good and evil are no longer so self-evident. We have to realise that each represents a *judgment*. In view of the fallibility of all human judgment, we cannot believe that we will always judge rightly. We might so easily be the victims of misjudgment. Nevertheless we have to make ethical decisions. The relativity of 'good' and 'evil' by no means signifies that these categories are invalid, or do not exist. Moral judgment is always present and carries with it characteristic psychological consequences. For moral evaluation is always founded upon the apparent certitudes of a moral code which pretends to know precisely what is good and what evil. But once we know how uncertain the foundation is, ethical decision becomes

a subjective, creative act. We can convince ourselves of its validity only *Deo concedente* – that is, there must be a spontaneous and decisive impulse on the part of the unconscious. Ethics itself, the decision between good and evil, is not affected by this impulse, only made more difficult for us. Nothing can spare us the torment of ethical decision. Nevertheless, harsh as it may sound, we must have the freedom in some circumstances to avoid the known moral good and do what is considered to be evil, if our ethical decision so requires. In other words, again: we must not succumb to either of the opposites.[7]

In other words, there is a duty or burden laid upon us, brought upon us, in a sense, by the development of consciousness within us. The imperative to judge – to make *our own* judgements; as Mary Midgley puts it, to 'discriminate' in the grey fog between our own perceptions of right and wrong, good and evil, in all our myriad decisions. But this can only be carried out with a proper validity from the base of a true relationship between the ego and the Self, as free as we can manage from the pressures of the shadow, the collective – the 'politically correct', and the grip of obsession.

We cannot and ought not to repudiate reason; but equally we must cling to the hope that instinct will hasten to our aid – in which case God is supporting us against God, as Job long ago understood.[8]

Towards the end of 'Late Thoughts' Jung exposes his own myth:

The message can then be understood as man's creative confrontation with the opposites and their synthesis in the self, the wholeness of his personality. The unavoidable internal contradictions in the image of a Creator-god can be reconciled in the unity and wholeness of the self as the *conjunctio oppositorum* of the alchemists or as a *unio mystica*. In the experience of the self it is no longer the opposites 'God' and 'man' that are reconciled, as it was before, but rather the opposites within the God-image itself. *That is the meaning of divine service, of the service which man can render to God, that light may emerge from the darkness, that the Creator may become conscious of His creation, and man conscious of himself.*

That is the goal, or one goal, which fits man meaningfully into the scheme of creation, and at the same time confers meaning upon it. It is an explanatory myth which has slowly taken shape within me in the course of the decades. It is a goal I can acknowledge and esteem, and which therefore satisfies me [my italics].

By virtue of his reflective faculties, man is raised out of the animal world, and by his mind he demonstrates that nature has put a high premium precisely upon the development of consciousness. Through consciousness he takes possession of nature by recognising the existence of the world and thus, as it were, confirming the Creator. The world becomes the phenomenal world, for without conscious reflection it would not be. If the Creator were conscious of Himself, He would not need conscious creatures; nor is it probable that the extremely indirect methods of creation, which squander millions of years upon the development of countless species and creatures, are the outcome of purposeful intention. Natural history tells us of a haphazard and casual transformation of species over hundreds of millions of years of devouring and being devoured. The biological and political history of man is an elaborate repetition of the same thing. But the history of the mind offers a different picture. Here the miracle of reflecting consciousness intervenes – the second cosmogony. The importance of consciousness is so great that one cannot help suspecting the element of *meaning* to be concealed somewhere within all the monstrous, apparently senseless biological turmoil, and that the road to its manifestation was ultimately found on the level of warm-blooded vertebrates possessed of a differentiated brain – found as if by chance, unintended and unforeseen, and yet somehow sensed, felt and groped for out of some dark urge.

I do not imagine that in my reflections on the meaning of man and his myth I have uttered a final truth, but I think that this is what can be said at the end of our aeon of the Fishes, and perhaps must be said in view of the coming aeon of Aquarius (the Water Bearer), who has a human figure …

Jung's myth is in a sense a creation myth, and with Christian undertones. But due care needs to be taken to understand 'creation' and 'God' within the whole context, the former taking full account of Darwin and the latter being closer to alchemy and the Tao than to orthodox Christianity.

I think Jung's myth comes close to my own, since I share both the view of an unconscious, yet developing All and man's significant role. I share a great love of the Chinese philosophers and the Tao but cannot accept the view which seems to be accepted by most Taoists of the error of asserting 'preferences'. On the contrary this seems to me our inescapable burden, our *karma*, and also our role, our *dharma*. I believe what the old philosopher was saying was not that we should put away preferences, but rather that the Tao was not easy, since preferences have their due.

As the *I Ching* propounds, life and nature are not static, and the seasons are not constant but forever changing. The cycle is never quite the same, year upon year. Man's progress advances, retreats, and hopefully advances a little more as errors are made, acknowledged and, again with hope, corrected. Wisdom can only be gained by experience. This is the course of individuation, which is never a straight path. It is also the fire of the alchemists, a slow distillation. It is also the course of philosophy. In the words of Iris Murdoch:

> There is a two-way movement in philosophy, a movement towards the building of elaborate theories, and a movement back towards the consideration of simple and obvious facts … Both these aspects of philosophy are necessary to it.[9]

And of Gabriel Marcel:

> The stage always remains to be set; in a sense everything starts from zero, and a philosopher is not worthy of the name unless he not only accepts but wills this harsh necessity.[10]

It is interesting to compare Jung's statement, 'we must cling to the hope that instinct will hasten to our aid – in which case God is supporting us against God', with Marcel's conception of hope:

> From this standpoint there is truly an intimate dialectical correlation between the optimism of technical progress and the philosophy of despair which seems inevitably to emerge from it – it is needless to insist on the examples offered by the world of to-day.
>
> It will perhaps be said: This optimism of technical progress is animated by great hope. How is hope in this sense to be reconciled with the ontological interpretation of hope?
>
> I believe it must be answered that, *speaking metaphysically, the only genuine hope is hope in what does not depend on ourselves,* hope springing from humility and not from pride. This brings us to the consideration of another aspect of the mystery – a mystery which, in the last analysis, is one and unique – on which I am endeavouring to throw some light.
>
> The metaphysical problem of pride – *hubris* – which was perceived by the Greeks and which has been one of the essential themes of Christian theology, seems to me to have been almost completely ignored by modern philosophers other than theologians. It has become a domain reserved for the moralist. Yet from my own standpoint it is an essential – if not the vital – question.

Both share the intuition that the only real hope, 'the hope against hope' that both man and Gaia need so badly, spring from the God-image in the depth of the psyche. Typologically both men were, at origin,[11] introspective intuitives – they looked inward and saw the broad picture, neglecting irrelevant detail. In any attempt to compare the writing of very different people, terminology is a great problem. One is reminded of old Professor Joad in the BBC wartime Brains Trust: 'It all depends what you *mean* by ...'

Marcel had a number of important concepts, 'being', and 'presence', and 'creative fidelity' in particular. These he explained as best he could, but others like the 'Idealists', whom he hated, are not so clear. Possibly he meant those philosophers who, in Iris Murdoch's words, 'impose a single theory which admits of no communication with or escape into rival theories' – those he regarded as having little conception of *hubris*.

> A presence is a reality; it is a kind of influx; it depends upon us to be permeable to this influx, but not, to tell the truth, to call it forth. Creative fidelity consists in maintaining ourselves actively in a permeable state; and there is a mysterious interchange between this free act and the gift granted in response to it ...
>
> When I say that a being is granted to me as a presence or as a being (it comes to the same, for he is not a being for me unless he is a presence), this means that I am unable to treat him as if he were merely placed in front of me; between him and me there arises a relationship which, in a sense, surpasses my awareness of him; he is not only before me, he is also within me – or, rather, these categories are transcended, they have no longer any meaning. The word influx conveys, though in a manner which is far too physical and spacial, the kind of interior accretion, of accretion from within, which comes into being as soon as presence is effective.[12]

I see huge parallels with Jung in Marcel. Marcel's 'creative fidelity' is an attitude very much akin to Jung's symbolic attitude as I have sought to describe it. An attitude of expectancy to symbolic communications from the Self which he called 'influx'. His sense of 'being' is a relationship with the Self in which we recognise in ourselves – and recognise in others – a totality and not simply ego or persona. It is a sadness to me that two of my favourite people never seem to have been aware of each other – Jung was only fourteen years older than Marcel. Marcel never seems to stray far from human nature into the wilderness of nature in its broad context, but his fidelity to the pre-eminence of experience is firm. He ends his autobiographical essay with the words:

> After all, the error of empiricism consists only in ignoring the part of invention and even of creative initiative involved in any genuine experience. It might also be said that its error is to take experience for granted and to ignore its mystery; whereas what is amazing and miraculous is that there should be experience at all. Does not the deepening of metaphysical knowledge consist essentially in the steps whereby experience, instead of evolving technics, turns inwards towards the realisation of itself?

While the introvert Marcel turns inwards in 'recollection' towards the centre of his being, the extravert Iris Murdoch turns outwards towards objects of 'attention' and towards the 'magnetic focus' of the Good. They have almost the same result, for although Marcel is more concerned with self-discovery and Murdoch with virtuous judgement and action, both are concerned to communicate as complete persons, Marcel *as one 'being' to another* and Murdoch with loving attention. Here is Marcel:

> And this brings us at last to recollection, for it is in recollection and this alone that this detachment [from experience] is accomplished. I am convinced, for my part, that no ontology – that is to say, no apprehension of ontological mystery in whatever degree – is possible except to a *being* [my italics] who is capable of recollecting himself, and of thus proving that he is not a living creature pure, and simple, a creature, that is to say, which is at the mercy of its life and without a hold upon it …
>
> The word means what it says – the act whereby I re-collect myself as a unity; but this hold, this grasp upon myself is also relaxation and abandon …
>
> This brings out the gap between my being and my life. I am not my life; and if I can judge my life – a fact I cannot deny without falling into a radical scepticism which is nothing other than despair – it is only on condition that I encounter myself within recollection beyond all possible judgment and, I would add, beyond all representation.[13]

I think that 'experience' here means simply immediate problems of life acting as a distraction. He uses it in with quite a different meaning in an earlier quotation. Here is Murdoch:

> However M of the example is an intelligent and well-intentioned person,

capable of self-criticism, capable of giving careful and just *attention* to an object which confronts her.[14]

I have used the word 'attention', which I borrow from Simone Weil, to express the idea of a just and loving gaze directed upon an individual reality. I believe this to be the characteristic and proper mark of the active moral agent.[15]

Virtue is *au fond* the same in the artist as in the good man in that it is a self-less attention to nature: something which is easy to name but very hard to achieve.[16]

I shall suggest that God was (or is) a *single perfect transcendent non-representational and necessarily real object of attention;* and I shall go on to suggest that moral philosophy should attempt to retain a central concept which has all these characteristics.[17]

Thus both stress the importance of communicating from a position involving the whole person and not merely the ego.

Murdoch's use of 'reality' threw me at first. When she uses it in connection with art, saying that Cézanne painted what was 'real' it suggests a photograph. 'Truth' would seem a better choice. Later she does admit a close connection between reality and truth and also transcendence. She uses reality primarily to distinguish from fantasy or illusion, but seems to me to denigrate fantasy unnecessarily with some loss of insight. What Cézanne painted was an inner image triggered by the landscape. This is not far from fantasy. What we admire is the truth that he saw: a transcendent version of reality, lovingly attended to.

Iris Murdoch draws heavily from art and beauty in her quest for God, for a personal myth that she can affirm, for that is surely her driving force, inspired by Plato, for whom I share her affection, more especially the works on Socrates.

The great artist sees his objects (and this is true whether they are sad, absurd, repulsive or even evil) in a light of justice and mercy. The direction of attention is, contrary to nature, outward, away from self which reduces all to a false unity, towards the great surprising variety of the world, and the ability so to direct attention is love.[18]

Again her language appears to need interpretation to me. By 'self' she means selfishness, the narcissistic ego or shadow, and by 'nature' *human* nature. By 'outward' I think she is intending a penetration of self-concern. What she actually describes is, in Jung's terms, an act of extraversion:

> In a sense therefore extraversion is a transfer of interest from subject to object. If it is extraversion of thinking, the subject thinks himself into the object. If it is an extraversion of feeling, he feels himself into it.   CW 6

Extraversion (and introversion) can be active or passive. In passive extraversion 'the object attracts the subject's interest of its own accord, even against his will'. Something like this must have happened to Murdoch's chosen artist, Cézanne, to provide the impetus to paint. That the subsequent scrupulous attention was enabled by love sounds true. In the next passage she scents her quarry:

> There is, however, something in the serious attempt to look compassionately at human things which automatically suggests that 'there is more than this'. The 'there is more than this', if it is not to be corrupted by some sort of quasi-theological finality, must remain *a very tiny spark of insight* [my italics], something with, as it were, a metaphysical position but no metaphysical form. But it seems to me that the spark is real, and that great art is evidence of its reality

Having finally found justification for the 'spark of insight' in art she can turn to morality. It is here that her focus on 'loving attention' is on a plane with Mary Midgley's discrimination in the grey fog between our own perceptions of right and wrong and Jung's appeal for instinct to come to our aid. It is almost four square with Marcel in his sense of 'being'. Whatever her scorn for the 'existentialist-behaviourists' (the Oxford philosophers), she is, like Marcel, a metaphysical existentialist. There are many paths to the centre, the Chinese *D'an T'ien*, where the spark of insight is more commonly felt to be located, including recollection or meditation, ritualistic exercises or simply a symbolic attitude or awareness such as enabled Murdoch's reaction to seeing a kestrel through the window, as recognised by Mary Midgley.

> There is nothing now but the kestrel. And when I return to thinking of the other matter it seems less important.[19]

She has been brought to a *higher level* in that synchronistic, symbolic moment. Looking down on her problems with the kestrel's vision of course they shrink to a human size.

Perhaps the following passage summarises her thinking:

> There is a magnetic centre. But it is easier to look at the converging edges than to look at the centre itself. We do not and probably cannot know, conceptualize, what it is like in the centre. It may be said that since we cannot see anything there why try to look. And is there not a danger of damaging our ability to focus on the sides? I think there is a sense in trying to look, though the occupation is perilous for reasons connected with masochism and other obscure devices of the psyche. The impulse to worship is deep and ambiguous and old. There are false suns, easier to gaze upon and far more comforting than the true one.[20]

In spite of her qualification, I am uneasy about the concept of worship, and Murdoch's 'attempt to look right away from the self towards a distant transcendent perfection'. Perfection is a sterile concept and to worship perfect goodness seems to me to detract from the mystery of consciousness and being, to 'explain them away'. Looking at the sun reminds me of the soul in the *Dream of Gerontius*: 'Take me away!' – but for reasons other than Newman had in mind. I would rather, in the spirit of *wu-wei*, do nothing and let the Tao work on me, expanding consciousness without forceful intent.

I could wish that Iris Murdoch, in these essays on the *Sovereignty of Good*, especially in her demolition of fantasy and day-dreaming, was not quite so 'worthy' and would let in some of the humour and compassion for human frailty so much more apparent in her novels. Even the most apparently trivial fantasy or image can prove enormously valuable when its import is suddenly recognised. The unconscious is no respecter of worthiness.

But I think Iris Murdoch's focus upon *loving attention* is an important contribution, as when she quotes from the old hymn of George Herbert:

> A servant with this clause
> Makes drudgery divine;
> Who sweeps a room as for Thy Laws
> Makes that and th' action fine.

There can be no doubt that any task in hand, if it is to be brought to a satisfying conclusion, particularly an art or craft, but also science and scholarship, needs to be done with the greatest of patient and gentle attention. Force and haste cannot profit, indeed sully the result. This is everyday wisdom that cannot be learned 'from books', but only from experience. It sometimes takes seventy years! This is the spirit of *wu-wei*; a truly spiritual exercise through which one may find ones centre which, for a Christian or Muslim, may be God. Attention of this kind usually follows a direct inspiration, but often leads to others. I think that her earlier words

> Virtue is *au fond* the same in the artist as in the good man in that it is a selfless attention to nature: something which is easy to name but very hard to achieve.[21]

were closest to the mark. It is close to Lao Tsu. Although Iris Murdoch takes her examples from painting, she regards literature as pre-eminently a source of inspiration, and has certainly practised what she preached.

It is interesting here to compare Murdoch's view of a focus to which we are drawn, as by a magnet, which is really the unreachable pole of Good, to the concluding thoughts of Teilhard de Chardin:

> … or yet again far from simply declining towards a catastrophe or senility – the human group is in fact turning, by arrangement and planetary convergence of all elemental terrestrial reflections, towards a second critical pole of reflection of a collective and higher order; towards a point beyond which (precisely because it is critical) we can see nothing directly, but a point through which we can nevertheless prognosticate the contact between thought, born of involution upon itself of the stuff of the universe, and that transcendent focus we call Omega, the principle which at one and the same time makes this involution irreversible and moves and collects it.[22]

De Chardin viewed mankind (or 'thinking-kind', which he called the noosphere) moving and converging towards a transcendent end or focus, which he called Omega. Although de Chardin did not regard this as inevitable, he did see creation – Gaia – evolving under the influence of consciousness towards a kind of distant but attainable harmony. I do not think Murdoch had this sense of evolution, which perhaps is due to her philosophic rather than scientific standpoint, but felt an impetus, perhaps a duty in mankind to move towards the Good. One feels a certain evolutionary optimism in her writing. Only Jung had the concept of

Creator and Creature being a complementary whole, evolving under the 'eyes' of consciousness.

I believe these four great thinkers to have been gripped by different, though related ideals emanating from their childhood experience of the ontological mystery – the mystery of being: Jung by consciousness, de Chardin by harmony, Murdoch by good and Marcel by being. For myself, the feel of the leaves in my hand was the trigger to my experience which has culminated in my interest in symbols, notably the diamond and the star. For Jung we can say it was the vision of the desecration of the cathedral. For the others we can only speculate.

NOTES

1 Gabriel Marcel, *The Philosophy of Existence*, Harvill Press (first published 1948).
2 C.G. Jung, *Memories, Dreams, Reflections*, Collins, 50.
3 Job, 40, 4.
4 C.G. Jung, *Answer to Job*, Routledge, 7.
5 *Ibid.*, xvii.
6 *Ibid.*, 61, 64.
7 *Memories, ibid.*, 312.
8 *Ibid.*, 314.
9 Iris Murdoch, *The Sovereignty of Good*, Routledge, 1.
10 Marcel, *ibid.*, 93.
11 Typology changes over the years according to experience. To those interested in astrology, Jung was a Leo and Marcel a Saggitarian.
12 Marcel, *ibid.*, 24.
13 *Ibid.*, 12.
14 Iris Murdoch *The Sovereignty of Good*, 17.
15 *Ibid.*, 34.
16 *Ibid.*, 41.
17 *Ibid.*, 85.
18 *Ibid.*, 100.
19 *Ibid.*
20 *Ibid.*
21 *Ibid.*, 41.
22 *The Phenomenon of Man*, Collins, 307.

3

The Mystery of Consciousness

'ALL PROFESSIONS are conspiracies against the laity,' once said Bernard Shaw and there is much truth in this. However I believe that, with the increasing use of the internet, and the mistakes made by all the professions, whether medicine, law or science, so much more in the public eye, the 'laity' are now far more educated and critical. In science in particular awareness of the importance of the environment has greatly increased. This is due in part to the excellent television programmes of Sir David Attenborough and not a little to the books of James Lovelock which, although requiring some knowledge and a great deal of concentration and retention, can nevertheless be digested now by a public which must have been much smaller when *Gaia* was first published. The fact that his books have been studied by philosophers, Mary Midgley in particular, speaks for itself. One of the great attractions of Lovelock's books is his lack of condescension; I think Bernard Shaw would have read them with pleasure. I am indebted to Lovelock for introducing me to Mary Midgley whose philosophic eye makes an important contribution to our understanding of our present condition. One place where we do seem to need an amalgam of viewpoints is that of consciousness.

There have been many attempts by academics of various disciplines to bring consciousness as a concept into their ambit of influence. Mary Midgley has devoted considerable attention to this:

> The modest fact that we are conscious is now agreed to constitute 'the problem of consciousness'.[1]

To the philosophers this forms a part of their branch of metaphysics and has been worried like a bone by many from the Greeks to Kant, Hegel and Dewey. To the best of my knowledge the 'problem' has defeated them. Scientists now have snatched at it. While some, continuing Gallileo, have tried to exclude from science everything subjective, others assert that only science can understand the real world:

> Recent years have seen an explosion of work in the sciences and humanities on science's last frontier, the problem of consciousness.[2]

After decimating this extraordinary statement, Midgley attempts a reconciliation. 'Neither apartheid nor conquest will work,' she says. She continues with an analysis of the concepts of mind and body:

> We have to avoid dividing ourselves up as Descartes did in the first place. *Things go wrong as soon as we start thinking about mind and body as if they were both objects* – that is, separate things in the world. The words *mind* and *body* do not name two separate kinds of stuff, nor two forms of a single stuff. The word *mind* is there to indicate something quite different – namely, ourselves as subjects, beings who *mind* about things. The two words name points of view – the inner and the outer. And these are aspects of the whole person, who is the unit mainly to be considered. Words like mind and body do not have to be the names of separate items. They, and the other many-sided words that we use for these topics – words such as *care, heart, spirit, sense* – are tools designed for particular kinds of work in the give-and-take of social life.

Although these words are many-sided, that is not to say that subjectivity can never be investigated in an objective sense. Many have attempted it and I will be looking in particular at the French existentialist Gabriel Marcel and the English philosopher – better known for her novels – Iris Murdoch. After all, the subjective content of another's mind constitutes an object, and can be considered objectively.

Just how it is to be investigated is, of course, important. I must speak from my own standpoint, as a chemist with an interest in Jungian psychology. Jung studied the minds, or psyches, of others all his life. He made observations about them as a scientist, especially the unconscious parts of the psyche and their influence, described them in his own words, and used his results to treat disorders in his patients: disorders not always ascribable only to the mind, but also to the body. What affects one must affect the other since, as Midgley points out, we are a whole, divisible only by convention, and throughout our lives operate (even the greatest philosophic thinkers) without much assistance from consciousness. Like all medicine, in its broadest sense of healing, psychology is an art as much as a science, and it is not very much use investigating from the bottom up (in Lovelock's words). Viruses have been studied for many years now but the common cold survives in regardless disdain. It will be objected that reductive research has proved enormously beneficial, at least in certain areas, and this is of course true, but (to my mind at least) research of this kind is not medicine, and mistakes continue to be made when the body (or mind) is used simply as a research tool and the whole person (or animal) is not treated empathetically by the healing arts. The best healers, it has always seemed to me, are not doctors but nurses, and I am sad to see the profession becoming more and more dominated by pharmacology. All that medicine can do is to help the mysterious healing forces on their way; and the best doctors know it.

As a scientist of the old-fashioned kind I have a love of research and the last thing I would wish is to be thought to denigrate it. From childhood up I have been filled with the wonder of science. Pasteur was my first hero. Leaving university I shuddered during the various interviews I attended with commercial organisations. To use my talents and all my hard work to make a cheaper plastic was simply not to be borne. Eventually I answered an advertisement for the government chemist department of a colony. A very leathery retired chemist asked me in a bored way why I thought I should enjoy such work. What on earth can I say, I wondered, without seeming some kind of sissy, or religious? At length I said rather timidly, 'I thought it might be more vocational.' To my relief his whole face lit up. 'Exactly,' he said and began to describe how the work would involve studying the features and the problems of the country for the benefit of its people. And so I left to analyse and map the soils of the country and later the nutrition of the coffee tree. At the age of twenty-five I was running my own laboratory and planning my own research. Not a Pasteur, perhaps, but doing real science nevertheless. Sadly, all that changed, but never my sense of wonder.

From the very first, observers of human behaviour recognised that the fundamental property of consciousness lay in an ability to discriminate values lying between opposites: black and white, male and female, full and empty, and this led ultimately to a moral discrimination. Good and evil were born and many myths developed which led to a variety of tribal mores. Among the earliest philosophers were the Chinese, who recognised, or postulated, a split in humanity from an unconscious state encountered in the remaining environment which they regarded as the All, the Tao. Allan Watts explains this impossible concept as well as anyone:

> If Tao signifies the order and course of nature, the question is, then, what *kind* of order? … It has rather the sense of *hsüan*, of that which is deep, dark, and mysterious prior to any distinction between order and disorder – that is, before any classification and naming of the features of the world.

> > The unnamed is heaven and earth's origin;
> > Naming is the mother of ten thousand things.
> > Whenever there is no desire (or, intention),
> > one beholds the mystery;
> > Whenever there is desire,
> > one beholds the manifestations;
> > These two have the same point of departure,
> > but differ (because of) the naming.
> > Their identity is *hsüan* –
> > *hsüan* beyond *hsüan*, all mystery's gate.
> >
> > *Lao-tzu* I, tr. Auct.

> The 'chaos' of *hsüan* is the nature of the world before any distinctions have been marked out and named, the wiggly Rorschach blot of nature. But as soon as even one distinction has been made, as between *yin* and *yang* or 0 and 1, all that we call the laws or principles of mathematics, physics, and biology follow of necessity, as has recently been demonstrated in the calculus system of G. Spencer Brown.[3]

'Naming' is the nature of consciousness and can be substituted for 'the manifestations' in the passage above. As soon as naming takes place polarities arise. As Watts explains: 'At the very roots of Chinese thinking and feeling there lies the principle of polarity, which is not to be confused with the ideas of opposition or conflict.'

> When all men recognise beauty, ugliness is conceived.
> When all men recognise goodness, evil is conceived.
> So existence supposes non-existence;
> The difficult is the complement of the easy;
> The short is the relative of the long;
> The high declines towards the low;
> Note and sound relate through resonance;

Before and after relate through following;
So the wise man needs no force to work,
And needs no words to teach.

Tao Te Ching, Chapter XI[4]

Contemplation of the polarities lies at the heart of much of Eastern religious exercise and ritual by which the adept enters into a harmonious relationship with the mysterious All, where such divisions do not exist – what many describe as a centring of the self. From the polarities arise the paradoxes dear to the Zen masters who would teach a postulant to contemplate a rock until he saw it move: movement in stillness. To Jung also the polarities were of paramount importance. He saw that irreconcilable conscious attitudes can lead to symbols arising from the unconscious, in dreams, fantasies or ideas, which can illuminate a problem from a new perspective, and introduce a whole new level of awareness.

At this point it is difficult to avoid words like *spirit*, *soul* and *love*. These are words, like also *myth* and *mystical*, which cause embarrassment to academics and the professions, and tend to be consigned to the bin of 'just rubbish' or 'not my kind of thing, you know'. This is partly because they evoke feelings and emotion or else savour of the magical or paranormal, all of which tend to interfere with a 'professional' attitude.

Mary Midgley does tackle heart and mind:

> The heart is the centre of concern, the mind is the centre of purpose or attention, and these cannot be dissociated. This does not prevent the mind from being the seat of thought, because thought in general is not information-handling or abstract calculation as computers do, but is the process of developing and articulating our perceptions and feelings …
>
> We are inclined to use words like 'heart' and 'feeling' to describe just a few selected sentiments which are somewhat detached from the practical business of living … as if non-romantic actions did not involve any feeling. But this cannot be right …
>
> We are in fact so constituted that we cannot act at all if feeling really fails.[5]

While reading this there came into my head – almost as a piece of nonsense – the old song from Kipling's poem 'Mandalay':

Come you back to Mandalay,
Where the old Flotilla lay:
Can't you 'ear their paddles chunkin' from Rangoon to Mandalay?

On the road to Mandalay,
Where the flyin'-fishes play,
An' the dawn comes up like thunder outer China 'crost the Bay

Said by many to be a geographical nonsense, but one of the most evocative of songs. It is also a consummate piece of rational articulation. The old soldier sighs for the lost magic of the East. He is seeing a series of visions. There is no 'road' as such. He sees the boats on the river, and the palm trees; then he is at the rail of the troopship watching the flying fishes, and then the dawn as the ship approaches Rangoon Bay, where the dawn does indeed come up out of China. The nostalgia evoked is overwhelming to one, like me, born in the East. Sadly, where are the gentle Burmese now? What I think brought this up was my reaction to Midgley's words, 'We are inclined to use words like "heart" and "feeling" to describe *just a few selected sentiments which are somewhat detached* from the practical business of living.' As she continues, 'This cannot be right.'

Apart from British understatement, Midgley's passage is fine, but it saddens me that it lies close to some of Jung's thinking while somehow skating off the edge. All of this is really the stuff of depth psychology, the study of conscious/unconscious interactions. This will be developed later.

I would like to end with a passage from Jung's autobiography, *Memories, Dreams, Reflections*, which shows that at the root of his lifetime study of the unconscious was an overwhelming sense of wonder at the mystery of consciousness itself:

To the very brink of the horizon we saw gigantic herds of animals: gazelle, antelope, gnu, zebra, warthog, and so on. Grazing, heads nodding, the herds moved forward like slow rivers. There was scarcely any sound save the melancholy cry of a bird of prey. This was the stillness of the eternal beginning, the world as it had always been, in the state of non-being; for until then no one had been present to know that it was this world. I walked away from my companions until I had put them out of sight, and savoured the feeling of being entirely alone. There I was now, the first human being to recognise that this was the world, but who did not know that in this moment he had first really created it.

There the cosmic meaning of consciousness became overwhelmingly clear to me. 'What nature leaves imperfect, the art perfects,' say the alchemists. Man, I, in an invisible act of creation put the, stamp of perfection on the world by giving it objective existence. This act we usually ascribe to the Creator alone, without considering that in so doing we view life as a machine calculated down to the last detail, which, along with the human psyche, runs on senselessly,, obeying foreknown and predetermined rules. In such a cheerless clockwork fantasy there is no drama of man, world, and God; there is no 'new day' leading to 'new shores' but only the dreariness of calculated processes. My old Pueblo friend came to my mind. He thought that the *raison d'être* of his pueblo had been to help their father, the sun, to cross the sky each day. I had envied him for the fullness of meaning in that belief, and had been looking about without hope for a myth of our own. Now I knew what it was, and knew even more: that man is indispensable for the completion of creation; that, in fact, he himself is the second creator of the world, who alone has given to the world its objective existence – without which, unheard, unseen, silently eating, giving birth, dying, heads nodding through hundreds of millions of years, it would have gone on in the profoundest night of non-being down to its unknown end. Human consciousness created objective existence and meaning, and man formed his indispensable place in the great process of being.[6]

Consciousness is perhaps the ultimate, ever-expanding, never-completed mystery born of the marriage of the diamond and the star.

The next chapter attempts through poetry to evoke some of the magic surrounding these great symbols and their relationship; and, by letting the imagination work on them and letting the thought processes run without any dumping of feelings or emotion, to catch a glimpse of the meaning lying dormant behind the polarities at the root of human consciousness.

NOTES

1 Mary Midgley, *Science and Poetry*, Routledge, 8.
2 *Ibid.,* 9, quoted from the Fourth Tucson Conference on Consciousness.
3 Alan Watts, *TAO: The Watercourse Way*, Penguin, 44.
4 *Ibid.,* 19.
5 Mary Midgley, *The Essential Mary Midgley*, Routledge, 200.
6 C.G. Jung, *Memories, Dreams, Reflections*, Collins and RKP, 240.

4

The Diamond and the Star in Poetry

Heaven and Earth

Twinkle, twinkle little star,
How I wonder what you are!
Far above the world so high,
Like a diamond in the sky!

THE NURSERY RHYME – actually a poem by Jane Taylor[1] – is perhaps a child's first experience of the mystery of the diamond, probably as yet unseen, a magical object taking its associative power from the visible mystery of the sky at night, perhaps the most numinous and awesome of the manifestations of nature. If the child is well-born, perhaps the mother will show her a ring on her finger with a small, clear stone breaking the firelight into a myriad colours, like a flashing rainbow – for the diamond has no light of its own. Perhaps her mother will carry her to the window and point to the numberless stars above, pulsing bravely against the black canopy of the unknown and unknowable, mirroring the sparkle of the diamond. But for the greater part of humanity the diamond is far beyond reach, the most precious jewel to be mined from the depths of the earth.

And so, perhaps, there is awakened in the child's young psyche some distant remembrance, creating a new wonder, reinforcing a lasting memory of that precious moment: a bright symbol of transformative power, a distillation of all that is most beautiful and most precious in the vastness and majesty of heaven and earth, uniting the highest with the lowest, the divine child of awakening consciousness.

Male and Female[2]

Star of royal beauty bright!

THE ASSOCIATION of the diamond with the star is paralleled throughout life and throughout society. As the star is the highest emblem of
honour, worn proudly upon the breasts of kings and the favourites
of the court and state,[3] so is the diamond the highest emblem of wealth
and power, shining from the crowns of kings and rajahs, and on the
fingers of their queens, and those who would emulate them. More than
that, both are symbolic of devotion: the star given as a symbol of devotion to the state, perhaps patriarchy, and the diamond as a symbol of a
man's devotion to a woman, whether she be a barmaid or Elizabeth
Taylor. So, as emblems of dress, the diamond and the star together unite
the male and the female: when the star is *yang* in this mode, the diamond
is *yin*.

There is in each case both a common element of expectation, and a
difference of time. The star is symbolic of past faithfulness and present
fealty, the diamond of present devotion and betrothal for the future. In
each case too there is a promise, and an initiation to a new condition of
life, a new maturity.

Yet is the diamond in its own right the supreme male gift, a male
emblem worn by a woman with receptive love and pride in her man, a
sparkling chip of luminous masculinity, clear as logic, sharp as the sword,
adamantine[4] – unbreakable and enduring to eternity. In this sense it is
supremely *yang*. The star, symbolic of male honour, light and consciousness, also has a feminine side. *Stella* and *astra* are feminine in gender. The
morning star was to the Native Americans a male divinity, the supreme
warrior, second only to the sun, and heralding the day, but the evening
star, 'Bright Star' and heralding the night, was a female divinity, the
mother of all beings.[5]

Fire and Water

When the stars threw down their spears
And watered heaven with their tears.[6]

AS WELL AS heaven and earth, male and female, the star and the diamond unite fire and water. Stars, which are in fact nuclear fires, are at their most moving and beautiful when seen reflected from the sea, just as the diamond finds its emotive home against the skin of a beautiful woman. Reflecting the fire of the stars, yet it is valued by its 'water'.[7] Like water, the power of emotion, it has the ability to split the purest white into the colours of the rainbow. And most valued is not the red or gold of the fire but rather blue – 'a blue of the first water'. In the depths of the stone, when it is pure of all flaws and inclusions, can be seen the blue of the sea, or the blue of the sky – the ozone layer of triatomic oxygen which, united to the burning hydrogen of the stars forms the water of the ocean.

... till at last
The long'd for dash of waves is heard, and wide
His luminous home of waters opens, bright
And tranquil, from whose floor the new-bathed stars
Emerge, and shine upon the Aral Sea.[8]

Good and Evil

Him the Almighty power
Hurled headlong flaming from th' etherial sky
With hideous ruin and combustion down
To bottomless perdition, there to dwell
In adamantine chains and penal fire
Who durst defy th' Omnipotent to arms.[9]

THE ASSOCIATION continues in myth and legend, and it is here, in this more adult world, that its shadow begins to emerge. Diamond derives etymologically from the Sanskrit *dyu*, meaning 'luminous being'. It partakes of much of the lore of jewelry. Shooting stars are symbolic of angels; when Lucifer fell, angelic light is said to have been embodied in stars and gems. Tibetan has the same symbol for diamond and thunderbolt.[10] All of this symbology suggests revealed knowledge or wisdom, 'bolts from the blue', something revealed suddenly and mysteriously of divine import, and it is noticeable that the source can be equally from the heights or the depths, from heaven or from earth; even from hell.

Go, and catch a falling star
Get with child a mandrake root
Tell me, where all past years are,
Or who cleft the Devil's foot.[11]

Jewels are the treasure of the earth, buried at the foot of the rainbow, symbolic of hidden wisdom, hard to find and closely guarded – usually by snakes, as most beautifully portrayed in *The Jungle Book* by the blind white cobra in the lost palace. Often snakes are given jewels as eyes, where the association of wisdom and its guardian is brought to a close proximity. The chthonic polarity is given sharpest expression in the stories, emanating widely from India, Arabia and Greece, that the diamond was once to be found in the jaws of serpents.

The shadow is not hard to see. Only the innocent, the pure in heart, can be trusted with it. As the 'heart's desire' it is the lust of thieves as much as the sacrificial emblem, the inspiration of envy as much as wonder, for love of wealth is the root of all evil.

Most significant, perhaps, is the association of the diamond with Lucifer, the 'light bearer', bound in 'adamantine chains', still shining, still peerless, in the bowels of the earth. Perhaps for Milton his 'penal fire' was the legendary furnace at the earth's core. As Lucifer is Satan, the shadow of Christ, so the star and diamond, as emblems of royalty are the begetters of jealousy, ambition, and bloodshed; for Bonny Prince Charlie, exiled in France, they symbolised the guilt of Culloden Moor:

> *Take away that star and garter –*
> *Hide them from my aching sight!*
> *Neither king not prince shall tempt me*
> *From my lonely room this night.*[12]

And the symbol of eternity so easily becomes the adamantine chain, the entrapment of the star-crossed lovers in the doom of an unhappy marriage. The sea of tranquillity becomes the storm-tossed ocean of marital strife. And the symbol of sacrificial devotion turns into the penury of alimony. In Leo Robin's immortal lyric:

> *Men grow cold as girls grow old,*
> *And we all lose our charms in the end.*
> *But, square shape or pear shape,*
> *These rocks don't lose their shape,*
> *Diamonds are a girl's best friend!*[13]

NOTES

1 'The Star', Jane Taylor, *Rhymes for the Nursery*.
2 'The Three Kings'.
3 The highest honours are also associated with the cross, as in Knight Grand Cross of the order of … (GB), Croix de la Legion d'Honneur (F). However the emblems are a star and ribbon or sash. The highest order in the British Raj was the Star of India.
4 From the Greek *adamas* – unconquerable.
5 *Larousse Encyclopedia of Mythology*, Hamlyn, 439. More particularly the Pawnees of Nebraska.
6 'Tyger, Tyger', William Blake.

 7 C.G. Jung, CW 12, 327 (218): 'Now it is – as I can hardly refrain from remarking – a curious "sport of nature" that the chief chemical constituent of the physical organism is carbon, which is characterised by four valences; also it is well known that the diamond is a carbon crystal. Carbon is black – coal, graphite – but the diamond is "purest water".'
 8 'Sohrab and Rustum', Matthew Arnold.
 9 *Paradise Lost*, John Milton.
10 *Ibid.* CW 9, Part 1, 636 (358).
11 'Go Catch a Falling Star', John Donne.
12 'Charles Edward at Versailles on the Anniversary of Culloden', W.E. Aytoun.
13 'Diamonds are a Girl's Best Friend', Leo Robin.

5

Towards a Symbolic Attitude

One picture is worth a thousand words
Chinese proverb

Certainly one of Jung's most important contributions to modern thought was his attempt to re-evaluate the role of the imaginal and symbolic in psychic life ...
J.J. Clarke, *Jung and Eastern Thought*

The Daoist future in the West is likely to take a different, albeit related, path from that of Buddhism. It is likely that the direction of this pathway will lie in its potential as an attitude of mind, rather than as a complete philosophy.
J.J. Clarke, *The Tao of the West*

The Symbol in Psychology

I HOPE THE LAST chapter will convey something of the wonder dormant, waiting to be sought and found, in the symbols of the diamond and star – for that is of course what they are: symbols, or, more strictly, images with a symbolic potential. Symbols are perhaps the foundation of Jung's psychology; its whole significance can be said to rely on an appreciation of their importance in and to the human personality and its development towards a goal of wholeness which he named 'individuation'. Perhaps the first stumbling block to an understanding of Jung is a failure to give symbols their due weight and value. This can never be achieved through the intellect alone, but requires the emotional content of a direct experience. The experience may be granted in many ways, through a sudden intuition, through reading a striking poem or listening to music or birdsong, as well as through some striking encounter

47

or event: 'surprised by joy', to borrow from C.S. Lewis, or hit by disappointment. A disappointment supplied the inspiration for this book. Emotion is always involved. And there must always be a hint of some meaning which is hidden, unknowable, unnameable, engendering wonder.

> A view which interprets the symbolic expression as the best possible formulation of a relatively *unknown* thing, which for that reason cannot be more clearly or characteristically represented is symbolic.[1]

A symbol always carries a stirring of the senses, a sense of wonder or even holiness, which Jung called the numinous and which distinguishes it from a mere sign or allegory, with which it is often confused. A sign, for example an exit sign, a policeman's helmet or a sergeant's stripes, portrays something *known* in Jung's definitions. It can be a universal ideogram, a consciously devised shortcut to the conscious understanding of a place or person, but often it carries the residue of an archaic symbol which has lost its numinous power.

> A symbol is alive only so long as it is pregnant with meaning.[2]

Joseph Campbell, in his book *Myths to Live By*, gives a most interesting quotation from the monk Thomas Merton:

> The true symbol [he states again], does not merely point to something else. It contains in itself a structure which awakens our consciousness to a new awareness of the inner meaning of life and of reality itself. A true symbol takes us to the center of the circle, not to another point on the circumference. It is by symbolism that man enters affectively and consciously into contact with his own deepest self, with other men, and with God. 'God is dead' … means, in fact, that 'symbols are dead'.

And yet sometimes the original numinosity of a commonplace sign can be restored in all its symbolic glory. Jung gives the example of the Christian cross, which can be taken as a sign of mere historical significance or imparted, as by the mystics, with an inexpressible form showing that for them it was a living symbol. This is true *par excellence* of the diamond and star. Their usefulness in a proper context as signs – in a pack of cards or on an epaulette – will never derogate from their value as symbols 'in another place'.

Whether a thing is a symbol or not depends chiefly on the attitude of the observing consciousness; for instance on whether it regards a given fact not merely as such but also as an expression of something unknown … *The attitude that takes a given phenomenon as symbolic may be called, for short, the symbolic attitude* [my italics]. It is only partially justified by the actual behaviour of things; for the rest it is the outcome of a definite view of the world which assigns meaning to events, whether great or small, and attaches to this meaning a greater value than to the bare facts.[3]

Thus to experience a symbol as numinous does require a certain attitude. Not, I would suggest, anything forced, but a readiness to expect the unexpected; the attitude in which, hopefully, one would start to read a poem; a willingness to indulge the unconscious as well as the conscious; something a little more sympathetic than an 'open mind'. I fear no reader would get very far into this volume with a cynical attitude.

The word *symbol* originates from the Greek *sun-ballein*, to throw together.[4] Within the psyche a symbol appears spontaneously, e.g. in dreams or fantasies, as a bridge between opposites, usually problems or opposed conditions which have been experienced as irreconcilable. When it does appear in such circumstances it carries with it a sense of release. It is as though the symbol acts as a gateway through a limbo to a higher perspective. Contents of the unconscious have been brought into the light of consciousness, recognised and valued, maybe forgiven, and brought to a condition of rest. It represents a small step along the path of pilgrimage.

Mattoon gives an example[5] of a woman who is unhappily married but feels she must stay in her unhappy state for fear of losing her children or facing poverty. Her problem seems insoluble. Then she has a dream of a tigress who is in pain but continues to care for her cubs. Looking into the symbol with the help of an analyst she suddenly realises that she too can 'fight like a tiger' but is not doing so. She can stand up for her rights and yet bear the pain of the marriage or divorce while continuing to look after her 'cubs'. A complete change of attitude has taken place. She is on a higher plane from which she can look down on her situation with a new perspective.

Jung described the operation of the symbol in this way as the 'transcendent function'.

But I would suggest also that if the attitude is present, and the attitude of the mother in my illustration is apt, the symbol may be manifested more as a gift of grace. The time may simply be ripe for a psychic transformation, an advance in consciousness; it may be quite trivial, or it may be of shattering significance.

The Symbol in Western Experience

WHAT IS THE CONTEXT of a symbol within the maelstrom of thoughts and feelings that make up our consciousness? In addressing the question of a fresh – better, perhaps, renewed – outlook with which I am concerned it seems to me that we need to get back to the roots of language, perception and interpretation. We must dig beyond metaphor, analogy and homology – back to the symbol.

I think few would now disagree that the immediate messages presented to us from the brain, e.g. from memories, which indeed most if not all thought requires, emerge into consciousness in a very basic language, the immediate language of our senses, or that these messages emerge from a level below or beyond a conscious threshold. Instinctive responses – removing a finger from a hot stove – seem to bypass the rational parts of the brain. However, messages involving understanding or feeling, whether sensations from the eyes, ears or fingers, or intuitions from the imponderable, must rely on memory for their contextual reference. They have been called reconstituted paradigms. We encounter, for example, a smell. A smell of tar and fish brings to mind a picture of a fishing port – normally one we remember. Conversely, a remembrance of a particular holiday can immediately bring back the remembered smell.

Since the eyes are so heavily involved, most messages take the form of images: a picture or pictorial sequence. But creative acts rely even more fundamentally on the engagement of the unconscious at a variety of levels. Perhaps the most difficult of the arts to understand or analyse is music. Sir Michael Tippett has some interesting thoughts:

> An artist like myself, who doesn't need to take drugs to enter a visionary world – and who is in a way hostile to the idea of such willed stimulus – has but to reach out with his 'metaphorical hand', and put the music down, because this music comes by instinct, a re-awakening inside the physical body, so that the stomach perhaps moves as the music moves. The movements of the stomach or any other part of the nervous system, in response to the

imagined music, is the somatic test of aesthetic validity within the combined psycho-somatic act of creation – just as sets of gradually acquired intellectual judgements of formal patternings, of taste, of values, are the tests of the conscious mind. When the music thus imagined is performed the process in the listener may be reversed – a psychosomatic response may be engendered that is analogous, or even similar, to the finally accepted optimum of the composer's tests ...

When I ask myself what music does really express, I find it difficult to define in words: not because words are another medium (for words, apart from their use in poetry, are the proper medium for definition), but because there is here something of the indefinable. However, there are certain things I can tell you about it. For instance, it must be concerned with the interior of my psyche – a very difficult term, because I don't even know where the psyche is, in the body or in the mind. But this is the vital thing. It's not about the sensations apprehended from the external world, but about the intimations, intuitions, dreams, fantasies, emotions, the feelings within ourselves. Insofar as poets use words and others use materials from the external world to express these feelings, I use sound. Whatever happens in the outside world – and I do take from the outside world – it must be in some way transmuted. Now this is a magical process which I cannot describe accurately. Nobody, indeed, knows what it is. And yet I have to operate it. Moreover, I operate it from a necessity so strong that I have only gradually come to accept it openly, and to understand that this obsessive process continues, persists, and may be so strong that I'm mystified finally as to what was in the outside world.[6]

Steven Rose, in his book *Lifelines*, analyses carefully the assumptions behind modern scientific attitudes and experimental results. Although focused on biology, much of its discussion is relevant to the whole stratum of scientific study from physics to molecular biology. At least the first part of his book, it seems to me, should be required reading in all sixth forms. While fully acknowledging the important position of reductive science, notably in most of physics and chemistry where a homeostatic situation can be assumed with known variables individually controlled, he draws attention to the many assumptions implied, often exposed with the passage of time as fallacious.

In his own speciality of biology he exposes its dangers, and the dangers of determinism in general, where the systems under examination are essentially dynamic, interactive in three dimensions, and the variables are numerous and always include the unknown and unforeseen. He stresses in particular the importance of the historical perspective of any experiment on a living system:

> [My main task] is rather to offer an alternative vision of living systems, a vision which recognizes the power and role of genes without subscribing to genetic determinism, and which recaptures an understanding of living organisms and their trajectories through time and space as lying at the centre of biology. It is these trajectories that I call *lifelines*. Far from being determined, or needing to invoke some non-material concept of free will to help us escape the determinist trap, it is in the nature of living systems to be radically indeterminate, to continually construct their – our – own futures, albeit in circumstances not of our own choosing.[7]

Many of the difficulties experienced by biologists (he points out) lie in particular in a lack of scrupulosity in expressing their work (or even understanding it) in terms of the assumptions made; in particular an insufficient understanding of the differences between metaphor, analogy and homology. When I went to school I learned the difference between a simile and a metaphor. Later I learned the difference between equality and equivalence. Later still I was introduced to functionality. This is no longer enough and we need to re-educate ourselves. The issue is so important that I will quote Rose's analysis in some detail:

> There are three ways in which 'resemblance' can occur, and everything revolves around which of the three applies in this case. Is the process I am studying in chicks best regarded as a metaphor for human memory, analogous to it or homologous to it? Biology uses, all three terms, but they are quite distinct in meaning and significance. In a *metaphor* we liken some process or phenomenon observed in one domain to a seemingly parallel process or phenomenon in a quite different domain ... Metaphors are not meant to imply identity of process or function, but rather they serve to cast an expected but helpful light on the phenomenon one is studying. None the less ... their seductive charm is highly dangerous.
>
> Like most such terms, analogy and homology have multiple meanings. In the context in which I am using them here, *analogy* implies a superficial resemblance between two phenomena, perhaps in terms, the function of a particular structure. Thus ... the heart can be regarded as a pump ... But analogies can also mislead – is it a help or a hindrance to regard the random access memory (RAM) in my computer as analogous to memory in chicks or humans?

By contrast, homology implies a deeper identity, derived from assumed common evolutionary origin. This assumption of a shared history implies common mechanisms. It is in this sense that the bones of the front feet of a horse may be regarded as homologous to those in the human hand.[8]

Rose distinguishes a view of the world as reducible to objects from a view of the world as a process in which occurrences crystallise, which seems more akin to non-Western philosophical traditions. By referring to non-Western traditions it seems as though he is beginning to recognise that some new approach, indeed vision, is needed. The difficulty lies in somehow marrying the reductionist traditions of science which continue to be so useful experimentally, with a view of the dynamism so evident in life and indeed the whole planetary system of which we are a part. But the particle and wave theories continue to live side by side in physics. Perhaps more remarkably their symbolic polarisation has been transcended by the more comprehensive theories of cosmologists.[9]

Another remarkable advance can be seen in chaos theory, emerging from the polarity of orderly mathematics and the disorder of the environment, epitomised by the weather. It seems to have been inspired by the beautiful images of those playing with fractals. As I write this I am very conscious of the disorderly nature of my thoughts! It will take a near miracle to bring them even to an acceptable order. In fact my thoughts often seem to fly in circles towards a strange attractor following the butterfly effect, beautifully demonstrated in the 2006 Royal Institution Christmas Lectures by Marcus du Sautoy. According to Wikipedia:

> The butterfly effect is a phrase that encapsulates the more technical notion of *sensitive dependence on initial conditions* in chaos theory. Small variations of the initial condition of a dynamical system may produce large variations in the long term behavior of the system ... The phrase refers to the idea that a butterfly's wings might create tiny changes in the atmosphere that ultimately cause a tornado to appear (or, for that matter, prevent a tornado from appearing). The flapping wing represents a small change in the initial condition of the system, which causes a chain of events leading to large-scale phenomena.

Du Sautoy caused a magnetised paintbrush to swing like a pendulum over a paper under which were three hidden magnets. The paintbrush swung in loops over the magnets, tracing a complex pattern and eventually aligning with one of them. Tiny variations in the original position of the brush caused it to end up in different positions impossible to predict. Wikipedia mentions an origin in literature in 1890 but I feel sure it was known to the ancient Chinese.

Another ascription for the wonderful symbol of the butterfly – that most exquisite of insects, fluttering in unpredictable directions – is the soul or *psyche* – and sometimes life itself. I am reminded of the biblical description of the spirit as the wind:

> The wind bloweth where it listeth and thou hearest the voice thereof, but knowest not whence it cometh and whither it goeth: so is every one that is born of the spirit.[10]

There is so much that is unpredictable in nature and in life. The butterfly effect so well demonstrates and symbolises their interconnectedness.

There is no reason why different experimental approaches should not continue to be used where they are most appropriate. Relativity has little place in building a house. This is an example of 'scale confusion' which is discussed later.[11] We need only to see its place, and our place, in the mysterious world we inhabit, and that inhabits us.

Scientists, perhaps all 'professionals' (should it be 'professors'?), abhor anything pertaining to the rather vague movement self-styled 'New Age' and the word they seem to have adopted, *holism* (more commonly used as *holistic*). (Rose refers to 'New Age propaganda'.) I find the polarisation of attitude and mutual distrust and dislike sad and unfortunate. There is a great deal of quackery involved in the practices of alternative medicine, but so there is in alopathic medicine, heavily disguised under academic correctness. Likewise there is much truth to be found on both sides. Some rapprochement is taking place. Acupuncture is more accepted and herbal remedies are more tolerated, as is Yoga. Even visualisation can be found in some hospitals. (I frequently use this unlikely technique in alleviating cramp and even muscular strains.) After all, the healing art progressed in a 'top-down' fashion as James Lovelock points out, subtitling his book 'The Practical Science of Planetary Medicine'. It should never be forgotten that healing is an art, and all art is more than canvas and paint or, indeed, pills. And, like it or not, we do live in a new age, the second millennium, possibly the last age of humankind.

Unfortunately rapprochement seems to have a dispersing effect. As those closest move towards each other, those further apart grow more hostile, eventually barricading themselves into a fortress of obstinacy from which the only escape seems death – of the intellect or body – or else a Pauline conversion. It seems to me that Richard Dawkins may soon find himself in such a situation. Another may be the American physicist Alan Guth. In front of me is a newspaper report of a hypothesis by the Cambridge physicist Neil Turok that the Big Bang proposed for the start of the known universe was not a unique event but one of many, possibly infinite events taking place between alternative universes. Guth began his demolition of this idea by showing a slide of a monkey. Turok's idea is one I have felt for many years to be plausible. It transcends the attractive but disproved continuously expanding universe of Fred Hoyle with the intuitively unattractive concept of a unique explosion. It is not so much that either is necessarily right or wrong (both are speculative, although many cosmologists now accept a hypothesis of 'eternal expansion' out of which universes are born) but that Guth so rudely demonstrated his entrenched position.

If I can persuade anyone of the overriding importance of maintaining an open mind, to see the terrible and stupidly unnecessary consequences of holding out doggedly to a closed view amounting to a dogma, I will have achieved something. I believe support may be found in the spectrum of comparisons or similarities to which Rose draws attention. But I want to get upstream of the metaphor –> analogy –> homology spectrum to what I believe is their origin, the symbol.

Metaphor is a linguistic device enabling something unknown to be explained in terms of something more familiar. But metaphor pervades the whole of language. This is well explained in Joel Primack and Nancy Ellen Abrams book *The View from the Centre of the Universe*: 'Science is both a consumer and creator of metaphors and is meaningless without *thousands* of them.'

> … metaphors are not just words or images that help describe a concept that already exists in the mind. Instead, metaphorical connection is the way the human brain understands anything abstract. The deepest metaphors are not optional or decorative: they're a kind of sense, like seeing or hearing, and much of what we consider to be reality can be perceived and experienced only through them. We understand almost everything that is not concrete

(even 'concrete' is a metaphor) in terms of something else. In short, the expansiveness of our metaphors determines the expansiveness of our reality. Those words or figures of speech that people call 'metaphors' are only the last flourish of expression of an unconscious connection called a conceptual metaphor, which is built into our thinking. For example, when we say that an unfeeling person is ice-cold, 'ice' is only the superficial metaphor; the underlying conceptual metaphor is 'affection is warmth.' Affection-warmth is a connection that every normal infant learns to make. [12]

I would like to suggest that the difference between a conceptual metaphor and a symbol is tenuous. A metaphor is essentially a linguistic device. 'Affection is warmth' represents a feeling. It is not language. A baby is not able to phrase it. Insofar as it can be recorded in its brain, it may be as an associated image of a mother's face. A subsequent event, long into the future, perhaps hearing a lullaby, may suddenly bring back the image, with the associated feeling as a symbol, from the unconscious. What I find particularly remarkable is the absence of the word *love*. What a baby learns, gazing into its mother's face in the warmth of her arms, is love, and there can be no greater symbol for love than that of the Madonna and child.

Devices of a similar nature are used to communicate where language cannot be used – for it is important to keep in mind that we communicate in many ways. As Tippett (above) says:

> An artist like myself ... has but to reach out with his 'metaphorical hand', and put the music down, because this music comes by instinct, a re-awakening inside the physical body, so that the stomach perhaps moves as the music moves.

Another non-linguistic form of communication is of course mathematics, pure and applied. Although mathematical scientists tend to work at the precision end of the spectrum, they resort to pictorial images in an effort to make the language of equations more comprehensible, and often the images convey information that they are unable to express (at least within a useful time frame) by equations. It is almost as though the metaphor is reversed. It is questionable, of course, to call a graph a metaphor for an equation. Yet it is not a homologue, not yet an analogue. It is the best way to communicate something which could not be understood (or at least with much more difficulty) in any other way.

Fractal images and computer models are quite good examples. An example is shown in Plate 6 of the free energy 'landscape' of the folding protein fragment known as 'the tryptophan zipper' made using a super-computer. The chemical formula is shown in the top right-hand corner. By folding, the energy state of the molecule is lowered so that it becomes more stable. Other arrangements are possible; thus there seem to be two 'valleys', but to change from one to the other the molecule must be given sufficient energy to rise over the unstable 'hills'. I have shown as a comparison the simpler, two-dimensional energy diagram of the two crystalline phases of carbon, diamond and graphite (Plate 14). Diamond has a slightly higher ground state energy than graphite, but to move from one state to the other requires a huge amount of energy to lift the element over the energy hill where the atoms are in a state of flux. The diagram shows that by using a catalyst the atoms can be rearranged with much less energy.

The protein fragment or polypeptide of Plate 6 is quite simple com-pared to even a small protein. It seems to serve in a protein to enable the folding process ('zipping') which is so important to these complex molecules which make up most of our bodies. So many of the reactions in life systems depend upon shape. It is very difficult to describe or even visualise the shapes of proteins and, without some means of commun-icating this, scientists are at a loss. It is easy to fall into error. Many tools are in use in hospitals and laboratories in ever greater efforts to 'see what is going on'. Many convert digital signals into images familiar to the eyes, so that it becomes difficult to remember that what one is seeing, for example in a CAT scan, is not actually a 'visual' image at all but a construct.

Plate 7 shows a remarkable picture of two folded states of the much larger enzyme protein phosphorilase involved in the body's energy utilisation. This is called a 'ribbon diagram', which is constructed from a study of the molecule's crystal structure. It used to be deduced from X-ray diffraction patterns pioneered by Dorothy Hodgkins and others, originally in the study of inorganic crystals. However the structures in living systems are so complicated that it requires a large computer to produce an image or model which gives a visual picture enabling the mind to grasp even a little of what is going on, which of course will always remain mysterious. What an image indeed.

Look on my works, ye Mighty, and despair![13]

I have compared this image with those involved in the elucidation of the structure of DNA, shown in Plate 8. There is a dispute over the recognition given to the discoverers, since it was Rosalind Franklin's deduction of the double helix pattern from her X-ray diffraction photo (left) that gave the final clue to Watson and Crick, who produced the structural model (right). It is evident that a finely trained mind is needed to understand the meaning of the former pattern, whereas the model gives an immediate understanding, not only of the structure itself but also of how it might operate in a living system.

It seems now that science is relying more and more on the use of images to convey meaning. Rose will not let us forget that these (bio-chemical) images are (in the main – some model *in vivo* processes in the body) derived from chemically manipulated 'dead' fragments which may be so far from their real, dynamic nature as to be misleading. (As I will show later, biology is not the only science where the use of images is increasing, as techniques of visualisation advance.)

The images have, in fact, Rose's character of metaphor, though, since they are non-linguistic, they are also symbols. The more familiar they become, the less their power to ignite the emotions, so that they die. One can only imagine the excitement of Watson and Crick when they inserted the last nucleotide into their model. Not all the emotion has been lost and the double helix still strikes us with awe. It remains, at least for many, a living symbol, though to those more familiar or less involved it may be merely a formula, rather like the Christian cross. Science in fact has always, and continues to progress through the inspiration afforded by symbols. What seems to be missing is the recognition that symbols operate through the psyche and carry their numinosity over into all the domains of human activity. A scientist is, first and foremost, human. I would like to quote a passage from Salley Vickers' novel *The Other Side of You*:

> 'By the way, something I wanted to ask. Have I got this right? Electrons have no material reality but are called into existence by us when we measure them!'
> 'Yes, that is right.'
> 'So in a sense they depend on us for their existence.'
> Hassid looked uneasy. 'Maybe, but …'
> 'But they don't figure, do they, in our reality until we set about looking for them?'

'I suppose not, yes.'
'And yet you would say they exist, wouldn't you?'
'Certainly they exist, Doctor.'
'Thank you, Hassid. That's helped me.'[14]

Symbols could be said to be a means by which something as yet only sought, or only a possibility, is called into existence. The Doctor's recognition that he loved gave it reality, and this arose from the symbol of the electron.

Although the differences of scientific approach among the different disciplines rest heavily on the history of their development, so well explained by Steven Rose, it must also be true to say that the differences lie as much in the psychology and *weltanschauung* of individual scientists. These differences manifest themselves not only in their particular disciplines but also in their attitude to life. Compare the biologist and geneticist Richard Dawkins:

> What is truly revolutionary about molecular biology in the post-Watson-Crick era is that it has become digital. There is no spirit-driven life force, no throbbing, heaving, pullulating, protoplasmic, mystic jelly. Life is just bytes and bytes and bytes of digital information ... Scientific beliefs are supported by evidence, and they get results. Myths and faiths are not and do not.[15]

with the chemist and earth scientist James Lovelock:

> Modern medicine recognizes the mind and body as part of a single system where the state of each can affect the health of the other. It may be true also in planetary medicine that our collective attitude towards the Earth affects and is affected by the health of the planet. Christian teaching has it that the body is the temple of the soul and that this alone is a good enough reason for leading a healthy life. I find myself looking on the Earth itself as a place for worship, with all life as its congregation. For me this is reason enough for doing everything that is in my power to sustain a healthy planet.[16]

I would suggest that Lovelock exhibits a symbolic attitude wholly absent in – even rejected by – Dawkins.

Rose also exposes the interventionist nature of experimental biology and the dilemmas it poses in the future development of moral precepts:

> We cannot escape the fact that interventionist biology, and above all physiology, is a science built on violence, on 'murdering to dissect', and that

hitherto there has been no alternative means of discovering the intimate molecular and cellular events that, at least on one level of description, constitute life itself. The reductive philosophy that has proved so seductive to biologists yet so hazardous in its consequences seems an almost inevitable product of this interventionist and necessarily violent methodology.

More than most sciences today, biology impinges directly on how we live.[17]

Our attitudes have never been so important.

At any moment the world, or the universe, is a simultaneity of patterns. The brain is capable of absorbing a view or a picture without translating it into some linear form, just as the nervous system could not operate through linear reasoning. Primack and Abrams emphasise the need for symbolic language in understanding the universe:

This modern icon of our gorgeous, undivided home planet (a photo of the Earth from space) shows the power of a new image to alter perceptions and attitudes. In this book we try to portray the universe in similarly meaningful pictures. We use the word 'picture' metaphorically, however. Any picture of the expanding universe can only be symbolic, because the universe can't directly be seen. No one can step outside of it to look at it, no one can see all times, and over 99 percent of its contents are invisible. Symbols are the way to grasp the universe. They free us from the limits of our five senses, which evolved to work in our earthly environment. Symbols let us see the universal with our mind, which doesn't have any obvious limits. And symbols are far easier to remember than a long, logical argument or a mathematical equation. Carl Jung wrote, 'A symbolic work is a perpetual challenge to our thoughts and feelings because even if we know what the symbols are, they do not refer to a given thing, like a sign, but are 'bridges thrown out towards an unseen shore.'[18]

They stress the need to connect with the universe in a cosmological myth, discussed more fully later. For this we need what I have been calling an open mind, which is perhaps a cliché. Their use of the expressions 'counter-intuitive' and 'suspension of disbelief' are better:

We need to connect to our universe. If all knowledge were intuitive, there would be no need for science. Relishing the counterintuitive can be liberating. The only human that has to be in the story is you. That's what it means to participate, but you can do this in your imagination. It is one thing to appreciate intellectually that time may be as this chapter described it, but quite another to suspend disbelief. Suspension of disbelief is the chance we give to any play or movie to touch us, independently of whether it's true. Suspension of disbelief is the chance we give ourselves to participate in the drama and the comedy.

The Symbol in Eastern Thought

WHILE FOR US in the West a symbolic attitude involves an act of will, the ancient Chinese lived their lives in a symbolic attitude. This is expressed through their writing, which does not use an alphabet but ideograms – shorthand images. Alan Watts, in his book *The Watercourse Way*, gives interesting diagrams showing the evolution of the final characters from their primitive origins, and also points to the increasing use of primitive ideograms in current usage as the world comes to terms with travellers with differing languages. Mobile phone 'texting' is now developing its own ideogrammatic shorthand. Unlike Japanese, or even ancient Egyptian, Chinese writing never developed into a linear form,[19] so that reading Chinese does not require a knowledge of the spoken language. In this respect it seems unique. This gives some insight into the writing of Lao-tsu.[20] As Alan Watts goes on to point out, the natural universe is not a linear system but 'involves an infinitude of variables acting simultaneously'. As Rose (above) said:

> We are dealing again with distinctions between object and surround, foreground and background. This latter way of conceptualizing the world is perhaps more akin to non-Western philosophical traditions

The linear thinking of Western culture has coped with the myriad variables of nature by division and subdivision, characterising and ordering, so that a science has developed which is not only disparate, but separated from and *opposed to* nature. We talk now of 'conquering nature' rather than seeing ourselves as a part of it – an insignificant part. As James Lovelock puts it in his *The Revenge of Gaia*:

> Science is a cosy, friendly club of specialists; it is proud and wonderfully productive but never certain and always hampered by the persistence of incomplete world views.[21]

Not so the ancient Chinese:

The beauty of Chinese calligraphy is thus the same beauty which we recognize in moving water, in foam, spray, eddies, and waves, as well as in clouds, flames, and weavings of smoke in sunlight. The Chinese call this kind of beauty the following of *li*, an ideogram which referred originally to the grain in jade and wood, and which Needham translates as 'organic pattern', although it is more generally understood as the 'reason' or 'principle' of things. *Li* is the pattern of behavior which comes about when one is in accord with the Tao, the watercourse of nature. The patterns of moving air are of the same character, and so the Chinese idea of elegance is expressed as *feng-liu,* the flowing of wind.[22]

Just as developing a symbolic attitude is a necessary step towards individuation in Jung's terms, so is it a step towards living in accord with the Tao of the ancient Chinese. Both represent the goal of a pilgrimage towards wholeness of mind and body and resonance with nature.

Above all, or perhaps at the root, the symbol represents a union of opposites. For Jung attention is focused on two pairs of opposites, epitomised by the sides or corners of the square, exhibited by the mandala, the Self symbol he found throughout Hindu and Buddhist mythology and in alchemy, and which can be found as far distant as Hawaii (Plates 2a-c).

The ancient Chinese equivalent lies perhaps in the eight ideograms of the *I Ching,* the great *Book of Changes* used in divination. These ideograms, formed of three full or broken lines called 'trigrams', are often shown in a circular form, illustrating the changing course of fate, but also exhibiting the pairs of opposites – heaven and earth, male and female, lake and mountain, wind and fire (Plate 9).

At the centre is the *yin-yang* symbol, *t'ai chi,* whose original meaning was the 'ridgepole', the line. According to Richard Wilhelm:

> With this line, which in itself represents oneness, duality comes into the world, for the line at the same time posits an above and a below, a right and a left, front and back – in a word, the world of opposites.[23]

As its name suggests, the *I Ching* is concerned not merely with opposites but with movement and change, reflected in the exercise or dance of *t'ai chi,** a conception at the root of Chinese philosophy, contrasting with the more static, meditative approach of India.

---

* I have kept to Wilhelm's spelling rather than the modern *tai ji.*

If we enquire as to the philosophy that pervades the book we can confine ourselves to a few basically important concepts. The underlying idea of the whole is the idea of change. It is related in the *Analects* that Confucius, standing by a river, said: 'Everything flows on and on like this river, without pause, day and night.' This expresses the idea of change. He who has perceived the meaning of change fixes his attention no longer on transitory individual things but on the immutable, eternal law at work in all change. This law is the Tao of Lao tsu, the course of things, the principle of the one in the many. That it may become manifest, a decision, a postulate, is necessary. This fundamental postulate is the 'great primal beginning' of all that exists, *t'ai chi* – in its original meaning the 'ridgepole'.[24]

Chungliang Al-Huang indeed visualises, and illustrates, the *t'ai chi* or *yin-yang* symbol as spinning (Plate 20b):

The line through the centre is wavy to perpetuate movement – like a flowing watercourse, not a permanently enforced division. Its meaning is in movement, in its dance of fluid change.[25]

Compare Rose (above):

it is in the nature of living systems to be radically indeterminate, to continually construct their – our – own futures, albeit in circumstances not of our own choosing

But this movement, this flow, bears no relationship to cause and effect, so essential to Western science and philosophy. The Chinese attitude is almost impossible for Western minds to grasp, let alone admire. The *I Ching* is, even now, generally regarded as superstitious nonsense. One wonders what Richard Dawkins would make of it:

What are all of us but self-reproducing robots?' he asked. 'We have been put together by our genes and what we do is roam the world looking for a way to sustain ourselves and ultimately produce another robot – a child.[26]

Or Wittgenstein:

The meaning of simple signs must be explained to us if we are to understand them.[27]

And:

Most of the propositions and questions of philosophers arise from our failure to understand the logic of our language.[28]

Wilhelm gives as good an exposition of the Chinese approach as may be possible:

> In order to understand what such a book is all about, it is imperative to cast off certain prejudices of the Western mind. It is a curious fact that such a gifted and intelligent people as the Chinese has never developed what we call science. Our science, however, is based upon the principle of causality, and causality is considered to be an axiomatic truth. But a great change in our standpoint is setting in. What Kant's *Critique of Pure Reason* failed to do, is being accomplished by modern physics. The axioms of causality are being shaken to their foundations: we know now that what we term natural laws are merely statistical truths and thus must necessarily allow for exceptions. We have not sufficiently taken into account as yet that we need the laboratory with its incisive restrictions in order to demonstrate the invariable validity of natural law. If we leave things to nature, we see a very different picture: every process is partially or totally interfered with by chance, so much so that under natural circumstances a course of events absolutely conforming to specific laws is almost an exception. The Chinese mind, as I see it at work in the *I Ching*, seems to be exclusively preoccupied with the chance aspect of events. What we call coincidence seems to be the chief concern of this peculiar mind, and what we worship as causality passes almost unnoticed.[29]

Wilhelm's reference to modern physics may refer to the uncertainty principle of Heisenberg, which developed out of the quantum theory of Niels Bohr, and possibly to the writings of Wolfgang Pauli, who corresponded with Jung. Some twenty-five years after Wilhelm published his translation Fritjof Capra wrote *The Tao of Physics*. The following passage helps to illuminate what Wilhelm had in mind:

> Quantum theory has thus demolished the classical concepts of solid objects and of strictly deterministic laws of nature. At the subatomic level, the solid material objects of classical physics dissolve into wave-like patterns of probabilities, and these patterns, ultimately, do not represent probabilities of things, but rather probabilities of interconnections. A careful analysis of the process of observation in atomic physics has shown that the subatomic particles have no meaning as isolated entities, but can only be understood as interconnections between the preparation of an experiment and the subsequent measurement. Quantum theory thus reveals a basic oneness of the universe. It shows that we cannot decompose the world into independently existing smallest units. As we penetrate into matter, nature does not show us

any isolated 'basic building blocks', but rather appears as a complicated web of relations between the various parts of the whole. These relations always include the observer in an essential way. The human observer constitutes the final link in the chain of observational processes, and the properties of any atomic object can only be understood in terms of the object's interaction with the observer. This means that the classical ideal of an objective description of nature is no longer valid. The Cartesian partition between the I and the world, between the observer and the observed, cannot be made when dealing with atomic matter. In atomic physics, we can never speak about nature without, at the same time, speaking about ourselves.[30]

The free energy graph of carbon at the foot of Plate 6 is a construct of 'old physics'. Under the 'new physics' of Capra, however high the energy barrier between two energy states there is always a finite probability (although very small) of particles crossing from one energy state to another and back again. This concept is essential to the understanding of semiconductors like transistors and computer chips. Without quantum theory computers (as we know them) would be science fiction. However the concept of 'probability' is impossible to visualise so we have the paradox in the upper figure of a construct of the new physics (the supercomputer) relying on old physics to map complex energy states.

Hence the symbol of the electron in Salley Vickers' novel. And it is really not so far from Rose's 'lifelines'. This speciality, at least, of modern science therefore focuses attention on the paradox of the 'moment' when all the interactions of nature, *including the observer*, can be said to merge. It is the pattern of the moment with which the *I Ching* is concerned, a moment in chronic time when even the vibrations of the diamond lattice or the solar reactions can be imagined as frozen.

In the words of Alan Watts:

> The point of view that underlies the *Book of Changes* is that instead of trying to understand events as relationships to past causes, it understands events by relation to their present pattern. In other words, it comprehends them by taking a total view of the organism and its environment instead of what we might call a linear view.[31]

The ancient Greeks recognised this aspect of time as *kairos*[32] (as distinct from *chronos*) and Jung gave it, or its properties, the name 'synchronicity'.

Jung had some interesting points to make on time in relation to dreams in a seminar published recently in *Jung History*.[33]

> For here we are facing such complex material, such a diversity and complexity of conditions, that no *unequivocal* causal connections can be maintained. Here the word *conditional* is much more appropriate …
>
> If one were to characterise the nature of dreams, one could say that they do not form a chronological series as in a, b, c, d … We rather have to suppose an unrecognizable center from which dreams emanate … The actual arrangement of dreams is a radial one: the dreams radiate from a center … In the final analysis they are arranged around a *center of meaning*.

Time has little or no meaning in relation to the unconscious. Often what appears to the conscious mind as a whole narrative takes place in an instant before waking.

It is interesting to compare the paradox of time described by Primack and Abrams:

> In eternal inflation, time doesn't mean anything on the large scale because it has no direction. What determines the difference between past and future in our ordinary experience is causality: past events cause present events, which in turn cause future events. Since everything in eternal inflation is flying away from almost everything else faster than the speed of light, which prevents even tiny regions from being in causal contact for more than an instant, there is no way to distinguish one time from another: everything is changing, yet overall nothing is changing. Eternal inflation is the ultimate steady state universe! Thus 'eternal' means not only that it never ends but also that there is no direction to time, and every time is the same as every other.

The above refers to a zone of 'eternal inflation' which is a hypothetical condition outside the present universe. It is perhaps the womb of all universes. However even within our own universe we are constrained within a region of spacetime defined by the speed of light. Primack and Abrams have pictured this as a double cone, past below and future above:

> The lightcone shows the strange implications of the speed of light. Perhaps most surprisingly, it shows that much of the information we will have in the future is already on its way towards us, but right now its not in our past or in our future. There is a third category: neither past nor future but 'outside our lightcone – perhaps 'elsewhen'.
>
> The region outside the lightcone represents most of the universe …[34]

Within the concept of spacetime, the slippery individual quantities of space and time compensate for each other, revealing an underlying unity that is not relative but absolute – or 'invariant'.[35]

The *I Ching* propounds a process of divination using coins or 'yarrow stalks' in which 'hexagrams' are discovered where one 'trigram' lies over another. Sixty-four such hexagrams are possible, each with its own message for the moment the oracle is consulted. When the trigram Heaven is over that of Earth there is indicated a time of stagnation, since the two trigrams are envisaged as separating. When, on the other hand, the situation is reversed with Earth over Heaven, the indication is a time of peace, since the symbols are said to be meeting in harmony.

Stagnation	*Peace*

Capra's work illustrates the profound differences that lie between the various branches of science, especially between biology and physics. Only now, in great part due to re-thinking by many scientist of differing disciplines such as the chemist James Lovelock, the biologist Steven Rose and cosmologists Joel Primack and Nancy Ellen Abrams, is some rapprochement taking place, under the duress imposed by dangers to the environment and the social and moral implications of advances in the life sciences. In Lovelock's words:

> James Hutton was a polymath who included medicine among his scientific qualifications. It was natural that he should, like some wise old barn owl floating above a meadow, have taken a holistic or 'top-down', physiological view of the Earth. To most scientists today however, James Hutton's idea of the Earth as a superorganism is contrary to conventional wisdom. The younger, expert sciences – whether microbiology or biogeochemistry – take a narrower, reductionist, or 'bottom-up' view, studying details and processes within their fields. They recognize that life on Earth affects its environment as well as adapting to it, but they have lost Hutton's great vision that saw life and its material environment as a single system.
>
> This is why I think that established science today is ill-suited to cope with the problems of global change. I am not suggesting we should cast aside our

vast store of scientific knowledge, still less that we abandon scientific method. But I do suggest we should look again at the evidence science has gathered, and see if the physiological view can better explain and predict what will happen next. In the 18th and 19th centuries, the science of physiology underpinned the empiricism of human medicine and enabled the physician to navigate a way to health through scientific uncertainties. Today, it can provide a basis for the empirical practice of planetary medicine and help us choose a path toward health for the whole system of Earth.[36]

Having got thus far into the book, and after researches into the *I Ching*, it seemed right to consult the oracle, following in the footsteps of Jung who did the same after writing his foreword to Wilhelm's translation. I felt my work was getting messy and I had not had a good day. But, looking out of the window, I saw the sun shining over the garden after some nasty weather. Perhaps things were taking a change for the better. I asked the question: 'How should I improve the book?' and received the answer:

This was Fire over Heaven, 'Possession in great measure', one of the most encouraging images expressed by the oracle, and it gave me a great boost. Provided I kept my modesty and obeyed the dictates of heaven, promoting the good, I would achieve success. In other words, it should not be regarded as an inflationary ego exercise, but as something above and beyond, with a spiritual dimension. The 'x' at the end of four of the lines showed them to be changing lines, each having a subsidiary meaning. The bottom line indicated that I was at the beginning and no blame would attach to mistakes. The next line was a reminder or admonition to offer my work to 'the ruler' – my Self as I read it – and not seek to regard the book as my own possession – my Ego. Next came an indication that I had powerful 'neighbours' (I regarded these as critics and competing authors), spelling some danger, and an admonition to keep straight in my purpose and not keep glancing around with envious looks. Finally the top line indicated good fortune, but that I should 'give honour to the sage', whom I understood as the authors to whom I owe so much, and perhaps to the book itself.

Of course the reflections are the result of my projections into the message, resting on my psychological state at or immediately following the 'moment'. But that is the purpose, and the power, of the oracle.

The changing lines imply that they can change into their opposites, giving rise to a new hexagram, which proved to be:

This was Earth over Water, 'The army', again a fascinating message indicating needs for discipline, good order and generalship, humane governance, and a clear aim. The results of this consultation were so pertinent that I wondered whether to include them in the book, again following Jung, but decided that this question was for another 'moment'.

And the diamond and star? As mentioned in the introduction they are a wonderful symbol for our planetary system of earth and sun.

Much more can be projected into the symbol if the imagination is allowed to wander, as I have tried to show in the previous chapter. Within the *I Ching* they would be seen as Earth and Heaven. As normally viewed, with the Sun above, it might appear that 'Stagnation' is represented, but this is to be forgetful of the antipodes under our feet. As the old hymn goes, 'The sun that bids us rest is waking our brethren 'neath the Western sky'. The system is in constant motion and change. Like the *I Ching*, the *t'ai chi*, the *yin-yang* symbol for the split of consciousness, itself points to the relative movement of the opposites: the decline of one means the renewal of the other, like the seasons, the succession of night and day, the waxing and waning of the moon – light growing from a seed in the darkness to its climax and fading, but never quite. Civilisations come and go. Over the aeons the diamond and star have seen it all.

NOTES

1 C.G. Jung, CW 6, 815.
2 *Ibid.,* CW 6, 816.
3 *Ibid.,* CW 6, 816.
4 Mattoon, *Jungian Psychology in Perspective,* 134 points to the German Sinnbild 'image of meaning' as being more direct. I find the Greek more apposite.
5 *Ibid.,* 138.
6 Sir Michael Tippett, 'Feelings of Inner Experience', from *How Does It Feel?* Thames & Hudson, 173, 174.
7 Steven Rose, *Lifelines,* Allen Lane, 7.
8 *Ibid.,* 33.
9 See Chapter 7.
10 John, 3, 8.
11 See Chapter 7.
12 Joel Primack and Nancy Ellen Abrams, *The View from the Centre of the Universe,* Fourth Estate, 13, 243.
13 'Ozymandias', Percy Bysshe Shelley.
14 Salley Vickers, *The Other Side of You,* Fourth Estate, 247. There is further discussion of electrons in 'Dreams and Myths'.
15 Richard Dawkins, collected quotations.
16 James Lovelock, *Gaia: The Practical Science of Planetary Medicine,* Gaia Books, 17.
17 Rose, *ibid.,* 4.
18 Primack *et al., ibid.,* 9.
19 It does, of course, involve a sequence of images to formulate a message.
20 There are many spellings for this Chinese sage, as Lao-tse, Lao Tzu.
21 James Lovelock, *The Revenge of Gaia,* Penguin, 5.
22 Alan Watts, *Tao, the Watercourse Way,* Pelican Books, 15.
23 Richard Wilhelm, *I Ching or Book of Changes,* RKF, Introduction.
24 *Ibid.*
25 Chungliang Al-Huang, *Quantum Soup,* Celestial Arts, 44.
26 Interview.
27 Wittgenstein, *Tractus Logica-Philsophica* 4.026.
28 *Ibid.*
29 Richard Wilhelm, *ibid.*
30 Fritjof Capra, *The Tao of Physics,* Flamingo, 78.
31 Alan Watts, *What is Tao?,* New World Library, 83.
32 Kairos, 'The Critical Moment', Liddell and Scott, Oxford, 1949.
33 Jung, C.G., *Jung History,* Fall 2006, Vol. 2, Issue 1, Philemon Foundation.
34 Primack *et al., ibid.,* 192.
35 *Ibid.,* 140.
36 James Lovelock, *Gaia: The Practical Science of Planetary Medicine,* Gaia Books, 10.

6

The Nature of the Unconscious

AT THIS POINT it seems necessary to review some further concepts of Jung, in particular his understanding of the nature of the psyche and the unconscious. Psychology, at least depth psychology, as opposed perhaps to behavioural analysis or prescriptive psychiatry, is an exploration of the unconscious. As such it falls between science (as it is generally regarded) and philosophy. Most scientists are only comfortable handling problems which they can verify in a purely objective sense. Theoretical physicists I happily except. Likewise philosophers try to confine themselves to the process of thought, which they consider to be rational and conscious. Neither discipline is at all comfortable to be considered subjective, or with the view that the subjective sense is capable of study in any meaningful way. To do so would be to fly in ever-decreasing circles like the infamous bird.

There are, of course, notable exceptions. Wolfgang Pauli and Fritjof Capra can be mentioned among scientists. But psychology has to grasp the nettle of subjectivity and grasp it tight:

This question, regarding the nature of the unconscious, brings with it the extraordinary intellectual difficulties with which the psychology of the unconscious confronts us. Such difficulties must inevitably arise whenever the mind launches forth boldly into the unknown and invisible. Our philosopher sets about it very cleverly, since, by his flat denial of the unconscious, he clears all complications out of his way at one sweep. A similar quandary faced the physicist of the old school, who believed exclusively in the wave theory of light and was then led to the discovery that there are phenomena which can be explained only by the particle theory. Happily, modern physics has shown the psychologist that it can cope with an apparent *contradictio in adiecto*. Encouraged by this example, the psychologist may be emboldened to tackle this controversial problem without having the feeling that he has dropped out of

71

the world of natural science altogether. It is not a question of his *asserting* anything, but of constructing a *model* which opens up a promising and useful field of inquiry. A model does not assert that something *is* so, it simply illustrates a particular mode of observation.[1]

But while Jung's theories on the *nature* of the psyche amount to a model, he never felt there was any doubt about the existence of the unconscious as an observable fact:

> I would like to clarify one aspect of the concept of the unconscious. The unconscious is not simply the unknown, it is rather the *unknown psychic;* and this we define on the one hand as all those things in us which, if they came to consciousness, would presumably differ in no respect from the known psychic contents, with the addition, on the other hand, of the psychoid system,[2] of which nothing is known directly. So defined, the unconscious depicts an extremely fluid state of affairs: everything of which I know, but of which I am not at the moment thinking; everything of which I was once conscious but have now forgotten; everything perceived by my senses, but not noted by my conscious mind; everything which, involuntarily and without paying attention to it, I feel, think, remember, want, and do; all the future things that are taking shape in me and will sometime come to consciousness: all this is the content of the unconscious.[3]

Jung quotes a passage from William James' *Varieties of Religious Experience* which is worth giving here also:

> Our whole past store of memories floats beyond this margin, ready at a touch to come in; and the entire mass of residual powers, impulses, and knowledges that constitute our empirical self stretches continuously beyond it. So vaguely drawn are the outlines between what is actual and what is only potential at any moment of our conscious life, that it is always hard to say of certain mental elements whether we are conscious of them or not.

Jung says this about the unconscious among the definitions in *Psychological Types*:

> To this extent, then, experience furnishes *points d'appui* for the assumption of unconscious contents. But it can tell us nothing about what might possibly be an unconscious content. It is idle to speculate about this, because the range of what could be an unconscious content is simply illimitable. What is the lowest limit of subliminal sense perception? Is there any way of measuring the scope and subtlety of unconscious associations? When is a forgotten content totally obliterated? To these questions there is no answer.

Our experience so far of the nature of unconscious contents permits us, however, to make one general classification. We can distinguish a *personal unconscious*, comprising all the acquisitions of personal life, everything forgotten, repressed, subliminally perceived, thought, felt. But, in addition to these personal unconscious contents, there are other contents which do not originate in personal acquisitions but in the inherited possibility of psychic functioning in general, i.e., in the inherited structure of the brain. These are the mythological associations, the motifs and images that can spring up anew anytime anywhere, independently of historical tradition or migration. I call these contents the *collective unconscious*. Just as conscious contents are engaged in a definite activity, so too are the unconscious contents, as experience confirms. And just as conscious psychic activity creates certain products, so unconscious psychic activity produces dreams, fantasies, etc. It is idle to speculate on how great a share consciousness has in dreams. A dream presents itself to us: we do not consciously create it. Conscious reproduction, or even the perception of it, certainly alters the dream in many ways, without, however, doing away with the basic fact of the unconscious source of creative activity.

The functional relation of the unconscious processes to consciousness may be described as compensatory, since experience shows that they bring to the surface the subliminal material that is constellated by the conscious situation, i.e., all those contents which could not be missing from the picture if everything were conscious.[4]

Jung recognised quite early in his life, during his time at the Burghölzli Institute, that the psyche was organised in groupings around certain emotionally-toned subjects which he called 'complexes'.[5] He realised this as a result of what has come to be known as his Association Experiment in which the times of reaction of a subject to certain words is noted, longer pauses being ascribed to emotional disturbance – touching a nerve, so to speak.

Strangely Jung does not define the word 'complex' among the other definitions in *Psychological Types*; but there is a good summary in the Appendix:

Whatever else may be taking place in the obscure recesses of the psyche – and there are notoriously many opinions about this – one thing is certain: it is the complexes (emotionally-toned contents having a certain amount of autonomy) which play the most important part here. The term 'autonomous complex' has often met with opposition, unjustifiably, it seems to me, because the active contents of the unconscious do behave in a way I cannot describe better than by the word 'autonomous.' The term is meant to indicate the capacity of the complexes to resist conscious intentions, and to come and go as they please. Judging by all we know about them, they are psychic entities which are outside the control of the conscious mind. They have been split off from consciousness and lead a separate existence in the dark realm of the unconscious, being at all times ready to hinder or reinforce the conscious functioning.

A deeper study of the complexes leads logically to the problem of their origin, and as to this a number of different theories are current. Theories apart, experience shows that complexes always contain something like a conflict, or at least are either the cause or the effect of a conflict. At any rate the characteristics of conflict – shock, upheaval, mental agony, inner strife – are peculiar to the complexes. They are the 'sore spots,' the *bêtes noires,* the 'skeletons in the cupboard' which we do not like to remember and still less to he reminded of by others, but which frequently come back to mind unbidden and in the most unwelcome fashion. They always contain memories, wishes, fears, duties, needs, or insights which somehow we can never really grapple with, and for this reason they constantly interfere with our conscious life in a disturbing and usually a harmful way.[6]

In dreams and fantasies, symbols emerge which more commonly arise from repressed complexes in the dreamer's personal unconscious. Normally symbols are not 'fixed'; that is to say their meaning varies with individuals and their circumstances, which is important in any attempt at interpreting dreams. There is always a temptation to project into the symbol a previously held meaning by the dreamer or interpreter. Sometimes they are common to his collective – the particular 'tribe' or community to which he belongs. But these may be associated with more general complexes common to all humanity and to which Jung gave names: the 'ego', the partly conscious concept we recognise as 'I', the 'shadow'; the parts of the ego we deny, the 'anima'; the repressed feminine values in a man, and its counterpart in a woman, the 'animus'. However there are symbols which are so powerful, and have been so widely recognised through the ages, that they must originate in what Jung calls the collective unconscious:

The collective unconscious contains the whole spiritual heritage of mankind's evolution, born anew in the brain structure of every individual. His conscious mind is an ephemeral phenomenon that accomplishes all provisional adaptations and orientations, for which reason one can best compare its function to orientation in space. The unconscious, on the other hand, is the source of the instinctual forces of the psyche and of the forms or categories that regulate them, namely the archetypes. All the most powerful ideas in history go back to archetypes. This is particularly true of religious ideas, but the central concepts of science, philosophy, and ethics are no exception to this rule. In their present form they are variants of archetypal ideas, created by consciously applying and adapting these ideas to reality. For it is the function of consciousness not only to recognize and assimilate the external world through the gateway of the senses, but to translate into visible reality the world within us.[7]

These symbols originating within the collective unconscious (there are very many such as the sun, moon, king, queen, square, circle and so on), Jung recognised as 'archetypal' in nature. Among such are symbols of great importance to individuation, to the achievement of personal wholeness, which he called 'Self Symbols', pertaining to the archetype of the self. An example is the mandala (Plates 2a-c).

Often regarded as an archetype, Jung is careful to define the self, among his definitions in *Psychological Types*, as an 'archetypal idea' representing the unity of the personality as a whole.

Jung's theory of complexes is fairly widely accepted as a working model and is being used, not least by police forces in lie detectors. He was in fact the first to use it in this way with the Swiss police. While working at the Burghölzli Institute as a young man he experimented with a form of test which he called the Association Experiment (mentioned above) in which he asked clients to give a response to a variety of carefully chosen and randomly arranged words, and timed their responses. An abnormally long pause he associated with a complex. In this way he was able to build up a certain picture of those complexes which seemed to be dominant in his client. It is still quite widely used by therapists, but timing has been replaced by electronic forms of detection in lie detectors. Steven Rose, in his book *Lifelines*, expresses some horror at the abuses of psychometric tests of which this, like his typology test (more commonly used in the Myers-Briggs version)[8] must be considered one. There are indeed some awful abuses, but these should not constitute a reason to dismiss them all.

It is not difficult to appreciate from personal experience that such complexes are 'triggered' constantly in everyday life and conversation. They give rise to emotional responses or affects. This is a normal interchange between the conscious and unconscious. Someone will make a remark to which we feel an immediate burst of anger or pity, or see a scene which 'strikes a deep chord'.

His ideas about archetypes and the collective unconscious are not so widely accepted, although they are also a useful model, which is how they should be viewed – not as some kind of mystic dogma to be consigned to the bin.[9]

Jung envisaged the unconscious as covering a spectrum extending from the instinctual realm, the red or infra-red to what he saw as a spiritual realm, which he thought of as blue, retaining the violet for the archetypes, where the spiritual and instinctive tended to mix:

> Just as in its lower reaches, the psyche loses itself in the organic material substrate, so in its upper reaches it resolves itself into a 'spiritual' form about which we know as little as we do about the functional basis of instinct.[10]

Instincts are involuntary. They do not involve any act of will. Jung thought of instincts as modes of action, but with an energy content or dynamism. But there is a need to account for phenomena of a more highly developed nature where the will becomes involved:

> But, over and above that, we also find in the unconscious qualities that are not individually acquired but are inherited, e.g., instincts as impulses to carry out actions from necessity, without conscious motivation. In this 'deeper' stratum we also find the *a-priori,* inborn forms of 'intuition,' namely the *archetypes* of perception and apprehension, which are the necessary *a-priori* determinants of all psychic processes. Just as his instincts compel man to a specifically human mode of existence, so the archetypes force his ways of perception and apprehension into specifically human patterns. The instincts and the archetypes together form the 'collective unconscious'. I call it collective because, unlike the personal unconscious, it is not made up of individual and more or less unique contents but those which are universal and of regular occurrence. Instinct is an essentially collective, i.e. universal, and regularly occurring phenomenon which has nothing to do with individuality. Archetypes have this quality in common with the instincts and are likewise collective phenomena.[11]

Therefore while instincts involve action, archetypes are essentially modes of apprehension. Although, like the instincts, inherited, universal and containing energy or dynamism, they do not have the force of compulsion but bring their images or ideas to the attention of the will. This can be accomplished in everyday activity by the process of intuition. Intuition, says Jung:

> is an unconscious process in that its result is the irruption into consciousness of an unconscious content, a sudden idea or 'hunch'. It resembles a process of perception, but unlike the conscious activity of the senses and introspection the perception is unconscious ... intuition is the unconscious purposive apprehension of a highly complicated situation ... But we should never forget that what we call complicated or even wonderful is not at all wonderful for Nature, but quite ordinary.[12]

Intuition[13] may intrude itself in many ways, though, to my mind, it appears to arise as an *idea* after the conscious rational mind has already begun its work of interpretation. The original intrusion is in a symbolic language, like Kekule's vision.[14] As a corollary, I wonder how far any process of thinking relies on intuition. It is, of course, intimately connected to memory, but the interconnections between memories are for the most part unconscious. The rational functions can direct the mind among the memories but the end result, the solution, can emerge quite suddenly, as though the mind was bored with the long-winded, step-by-step, cause-and-effect process of directed thought and short-circuited it altogether.

As memories have long been considered subject to extensive blocking to avoid flooding of the brain with unwanted images, so it has been postulated by Bergson and others that a similar effect is present to shield the brain from too much access to the unconscious. This type of filtering, so the hypothesis states, can be broken down, for example through the use of drugs like LSD and in neurological disorders like schizophrenia. Aldous Huxley wrote, after experimenting with mescaline:

> I find myself agreeing with the eminent Cambridge philosopher, Dr C.D. Broad, that we should do well to consider much more seriously than we have hitherto been inclined to do the type of theory which Bergson put forward in connection with memory and sense perception. The suggestion is that the function of the brain and nervous system and sense organs is in the main

eliminative and not productive. Each person is at each moment capable of remembering all that has ever happened to him and of perceiving everything that is happening everywhere in the universe. The function of the brain and nervous system is to protect us from being overwhelmed and confused by this mass of largely useless and irrelevant knowledge, by shutting out most of what we should otherwise perceive or remember at any moment, and leaving only that very small and special selection which is likely to be practically useful.[15]

This idea of filtering seems to me very plausible. Intuition can then be considered as a penetration of the filter similar to the process of seeking an old, unused memory. However this must obviously remain a working model for as long as it is useful. Neuro-biologists like Steven Rose are making rapid advances in elucidating the mysteries of memory and will have much to say. I find it sad that Rose (whom I so much admire) should appear to criticise Jung unfairly in *The Making of Memory*:

> The psychoanalyst Jung based his theory of mind in part on the claim that such collective memories were racial and had become deeply inscribed in our biological as well as cultural inheritance. I of course mean nothing of the sort here; rather I am talking about mechanisms of retention and trans- mission – the sharing and collectivisation of memories.[16]

The passage appears to suggest by the word 'racial', rather than simply 'historical', that Jung's theory of the collective unconscious was somehow confined to the perpetuation of prejudice rather than 'the necessary *a-priori* determinants of all psychic processes'; and furthermore, by the phrase 'nothing of the sort', that it was scientifically unsound – both suggestions being far from true. There can be no doubt that instincts (modes of action) are part of our biological inheritance and little reason to doubt that modes of apprehension can be inherited in the same way, rather the contrary.

This may be another example of scale confusion, or a confusion of 'levels of organisation' as Rose puts it. Just as cosmologists talk of confusion among scales of dimensions, so biologists talk of similar con- fusion among levels:

> To put it formally, we live in a material world which is an ontological unity, but which we approach with epistemological diversity. Biology, and the life processes it studies, will not conform to the proud manifesto of physics that

the task of science is to reduce all accounts of the world to unitary theories of everything. Physics' claim will not work, and it is positively harmful to our understanding of living processes.
Different scientific disciplines, from the social to the subatomic sciences, deal with different levels of organization of matter. The divisions between levels are, however, confused. In part they are ontological, and relate to scale and complexity, in which successive levels are nested one within another. Thus atoms are less complex than molecules, molecules than cells, cells than organisms, and organisms than populations and ecosystems. So at each level different organizing relations appear, and different types of description and explanation are required. Hence each level appears as a holon – integrating levels below it, but merely a subset of the levels above. In this sense, levels are fundamentally irreducible; ecology cannot be reduced to genetics, nor biochemistry to chemistry. However, to some extent – and this is where the confusion enters – the levels are epistemological, relating to different ways of knowing the world, each in turn the contingent product of its own discipline's history. The relationship between such epistemological levels (between biochemistry and physiology, say) is best described in the metaphor of translation. Thus the physiological language of contraction of the frog muscle can be translated into the biochemical language of the sliding filaments of actin and myosin. Problems arise when one attempts to apply concepts and terms applicable at one level to phenomena on another level. Thus people may be gay or violent or schizophrenic or selfish, but brains or genes cannot be in anything other than a metaphorical sense.[17]

And, of course, mixing metaphors leads to absurdity. Clarke writes about Jung's Unconscious:

> The emphasis on wholeness and unity as the goal of psychic transformation – the individuation process – can lead the unwary to suppose that for Jung the human psyche is ideally some kind of homogeneous substance, and that the splitting of the self is always pathological. Quite the contrary, Jung saw the self as a multiplicity of related elements, a dynamic system of psychic processes that seeks, not the elimination of multiplicity, but the bringing of this multiplicity into awareness and into balance. His view on this matter is close to that of modern systems theory which views wholes in terms of the homeostatic balance between competing forces, such as, for example, in an ecosystem or an organism.
>
> In *The Secret of the Golden Flower* the quest for spiritual enlightenment has, as Jung read it, a similar structure.[18]

Finally a little more perhaps needs to be said about the 'shadow'. This is envisaged as the 'dark' side of man and has connotations of evil and shame. Insofar as most repressed memories tend to be of shameful events, it is apt. However, since it can also be said to represent everything not lit by the light of consciousness, it is sometimes used to represent the whole content of the personal unconscious. In this aspect not everything can be said to be evil or even shameful. Many events experienced as shameful at one time can be recognised later on as merely actions forbidden in childhood or by a transient societal mores, e.g. of 'respectability'.

On page 151 of *Psychology and Alchemy* (CW 12) is shown a Kabbalist picture of 'Heaven fertilising Earth'. Earth is shown as a woman called by Jung 'the dark and dreaded maternal womb'. Although somewhat strange, it is a representation of the divine marriage, the *conjunctio* depicted in Plate 3b. On an individual level, the *conjunctio* symbolises the complete or individuated person or his conception. The paternalistic religions often regarded the woman, Eve, as the source of original sin. In alchemy she was closer to the *prima materia*. The light of heaven was needed to conceive the whole person. She might represent carbon, the shady womb of life. But I prefer to hold the image of the *Diamond Body* fertilised by the Star to conceive the Tao.

NOTES

1 C.G. Jung, CW 8, 381.
2 The automatic bodily functions.
3 *Ibid.*, 382.
4 C.G. Jung, CW 6, 842.
5 Jung did not originate the word, devised by Ziehen.
6 Paragraphs 923, 924.
7 C.G. Jung, CW 8, 342. (Compare Tippett in the previous chapter.)
8 Discussed in Chapter 8.
9 See passage quoted above, and Rose: 16 below.
10 C.G. Jung, CW 8, 380.

11 *Ibid.*, 270.
12 *Ibid.*, 269.
13 Further discussed in Chapter 8.
14 See Introduction. (Compare Tippett's 'movements of the stomach', p50, 56.)
15 Quoted in Joseph Campbell, *Myths to Live By*, Viking, 264.
16 Steven Rose, *The Making of Memory*, Bantam, 62.
17 Steven Rose, *Lifelines*, Penguin, 304, 305.
18 J.J. Clarke, *Jung and Eastern Thought: A Dialogue with the Orient*, Routledge, 85.

7

Dreams and Myths

PERHAPS THE MOST common form of unconscious intrusion, or most widely recognised, is via dreams, which of course are extensively studied during an analysis.

Dreams have long been revered as messages from the beyond – beyond certainly the dreamer's waking consciousness. That the images presented upon waking, sometimes fleeting, sometimes striking or fearful, emanate from an unconscious region of the psyche must be generally accepted.

Freud was the first to treat dreams seriously as of psychological importance. Being grounded in Jung's writings, I must admit to some ignorance of (Freudian) psychoanalysis. My understanding is that Freud concentrated on neuroses arising out of childhood repressions, particularly having a sexual grounding, and regarded dreams as representing wish fulfilment, the language being disguised. However there have been considerable changes in outlook and I am greatly indebted to Margaret Arden's *Midwifery of the Soul* for some explanation of these. In particular wish fulfilment and disguise no longer seem to be adhered to:

> However, it is no longer possible to hold his other view, that this subtle process has as its aim the discharge of repressed wishes. The only sensible view is that waking and dreaming consciousness are complementary forms of mental activity ...
>
> The aim of psychoanalysis is the recovery of a sense of belonging in the world through getting in touch with aspects of the self that were disowned in childhood. Paying attention to our dreams is one way of reversing the splitting processes that fragment society. In this way psychoanalysis connects us with the pre-enlightenment state of participation ...
>
> In conclusion, the essentials of psychoanalysis are reaffirmed by these new developments in science. Analysts have been subtly influenced by the changes

in the world over the decades so that modern practice bears little relationship to the supposedly value-free psychoanalysis that Freud first conceived. Repression and disguise are not required to the extent they were in Freud's Vienna. The changes in society are reflected in patients' dreams, which always have a cultural dimension. The growth in self-awareness that the Freudian revolution produced has resulted in much greater openness to self-understanding.[1]

To Jung, for whom the unconscious was so much wider and more profound, they bring to the dreamer's attention matters needing attention for one reason or another. They arrive usually from an unbalanced psychic situation, e.g. when too much energy is being devoted to one cause or another at the expense of something more psychically important, where they exert a compensating function. This is not always the case. Self images, emanating from a deep level of the unconscious, can direct or encourage the dreamer towards new psychic states. They can also be prophetic.

The language, far from being in some form of disguise, was to Jung strictly pertinent, albeit arrayed in imagery requiring exercise of the imagination to understand. The images, generally personal and currently relevant, may originate from any depth and often carry more than one meaning from different levels. Sometimes the meaning may be obscure for months or even years, and then suddenly the meaning is realised; maybe the circumstances have repeated themselves, which is not uncommon. Even when not fully understood, the compensatory effect can take place at an unconscious level. They are often ambiguous and sometimes baffle all attempts at an understanding. The imagery knows nothing of shame or prurience and some of my dreams are beyond publication.

The retention of dream images upon waking demands some skill and practice, and unless the discipline is continued few dreams are remembered. During a long period of analysis I kept a small recorder by my bed and would mumble the dreams as I woke, often in the night. These would be transcribed later, but not too late or the original image would have vanished. This practice produced, if anything, rather too many dreams, and many would never be fully attended to. I painted the important ones into a notebook which serves as a fascinating diary. But dreams are always there to accept.

Self images appear infrequently, but sometimes, seemingly as an encouragement, near the beginning of an analysis. One such is shown in Plate 10: a vision of a magical island city in the distance, reminiscent of St. Augustine's City of God, reachable by a journey over the sea of the unconscious. The shadowy area to the right may be a reminder of work still to be done in the lifetime remaining.

The image of the magical city returns from time to time. The following dream, some five years later, gives an idea of the many different messages, at different levels, which can emanate from a single dream image.

I dreamed of a strange insect, cartoon-like, with spidery arms and legs and wearing a red hard hat (Plate 11). This puzzled me for some time until I recalled I had been reading about Psyche's labours, imposed by Aphrodite, where the ants helped her to sort out numberless grains into piles. I had been obsessing over an apparently insoluble problem at work which seemed to have no solution. On awaking after the dream, I realised that the problem had been solved in the night. Ever since I have let my little ants in the unconscious mysteriously sort out difficult problems by 'sleeping on them'.

The Greek myths provide a huge font of images and ideas in at least the European unconscious Other sources are alchemy, as Kekule's Uroboros (Plate 4a) and Christianity, as Murdoch's angels (discussed later). In America images from the native mythology of North, Central and South America are also often encountered.

On returning to my notebook, I remember that the image had other significant features. I called the figure 'Alpha', which I think he must have told me in the dream. He is dressed in motley like a court jester and carries a sword in his right hand and a mirror in his left. Court jesters, of course, were the only beings permitted to ridicule and show up the faults of their royal masters. The sword resembles a Malay *kris*, which is flame-shaped and mirrors the winding path on which he stands. Malaya was the land of my childhood. In the distance is a futuristic city towards which the path leads. In the sky are a flying fish and a star, which I see as helpers. Fish for me are denizens of the unconscious and the star gives light in the darkness – this time the morning star in the twilight. The huge significance of twilight zones will be discussed later. On the right of the path is a tower which seems to pierce two ring-like floors. This may symbolise a phallus or act of procreation. (I have two children.) On the left are igloo-like buildings resembling eggs. The way to the city, the end of the pilgrimage, or of life itself, lies between them and past them. I do not think the dream needs further explanation though what I see in it are my own projections.

I find it exasperating that Iris Murdoch, one of the most individuated people I have run across, never seemed to have studied Jung. There is no mention of him in Peter Conradi's biography.[2] The biography watches her grow from a promiscuous, rather strident youth into the most un-assuming yet erudite old age in a very blessed relationship. Andrew Harvey's[3] recollection is worth quoting:

> [He] spoke for many of Iris's friends in seeing her as a sage (no saint) who gave all her friends unstinting, patient and non-judgemental support, making them feel loved, blessed, accepted, unique. He noted the reserve which marked her natural dignity. She had no need to impress or prove anything, was an astonishing example of how to wear fame and assume the dignity of an elder, never for one second the *grande dame*. Her natural radiation stemmed from a powerful, peaceful, gentle wisdom, her journey an increasingly wide embrace from an increasingly private centre. She had '360° mindfulness-awareness'. He intuited the 'work' she had done on herself, helped him midwife his own mystical experiences, recorded a three-hour interview with him about Buddhism for the American journal *Tricycle*, acted as guarantor of what he was coming to understand, showing him how to *be*.[4]

That she had a continuous relationship with the unconscious is clear. I have mentioned her vision of the kestrel from her window. Although much of her characterisation was taken from life observation, her imagination owes much to her inner life. Conradi mentions some dreams of which the following may be mentioned:

> In April 1982 she dreamt of a 'rather ridiculous-looking' yet beautiful Tibetan wearing a sort of dhoti with European coat and bowler hat. Unsure if she was allowed to speak to him, she felt healed and thrilled when he touched her back, experiencing strong desire when she leant her face gently against his and he advised her not to kiss him. After darkness and sleep she 'awoke' in Oxford High Street, where she invited the holy man to drink in a pub: 'a mistake'. The very rude landlord would not serve drinks. There were now eight or twelve people, 'very upset', who ate a meal together. A woman sitting beside Iris said, 'He pardoned you.' Iris replied, 'Yes,' but was sorry that he

had evidently spoken about her to a third party. After some confusion and 'a sort of pink substance smeared on my face (some ritual)', she felt excluded and that she must go away.

At this point she awoke (about 6.30am) and tried to continue the dream in drowsy waking thoughts. It went on as a conversation. She wanted to find out 'what he said to me', which turned out to be: 'Give up drinking. Live a quiet orderly life. Bring *peace* and *order* into your life. Give up certain thoughts, send them quietly away. When you feel your clutching craving hands holding onto something, gently detach them. Sit, kneel, or sometimes lie on your face in a quiet room. Have flowers in the room. Love the visible world.' She asked him about penance. 'Penance? Think in that way if you like, but not with intensity.' 'Can I see you sometimes?' 'I am nobody, you must give me up too.' A feeling of grace, of a door opening, accompanied these half-waking thoughts.[5]

The strange Tibetan reminds me of Alpha and must be a Self image. She is uncertain how to treat him but recognises his holiness during her subsequent meditation, what Jung would term 'active imagination'.

Another important dream follows, which she used thirty years later in one of her novels:

She saw in a garden two allegorical figures of birds of prey who transmuted into angels, be-winged and with golden hair. They came down from their pillars, passing her by. I follow them, a little afraid, and call after them. 'Can I ask you one question: Is there a God?' They reply 'Yes,' and disappear round a corner. I follow them and find myself alone in a gravel walk by the side of a building. Then I hear the footsteps approaching of someone whom I know to be the Christ. Filled with an indescribable terror and sense of abasement I fall on my face. The footsteps pass me and I hear a voice say: 'Ite' – which I take in the dream to mean 'Come' in Greek. I dare not look up. She heard another person approaching with a rustling dress. This person – the Virgin Mary – stopped beside her and put her hand on Iris's shoulder: 'The burden of terror is lifted a little and I say "Forgive me."' She replies to the effect that 'Your sins are forgiven.' The Virgin passed on, Iris keeping her face hidden.[6]

Iris Murdoch struggled with Christianity, which she could never wholly discard, although happier in Platonism and later Buddhism. There was probably guilt amid the struggle. In the dream she did not follow the old call to discipleship, but the terror of relinquishing the God of her childhood was real. Yet her 'sins' were forgiven. I can empathise with these feelings. One can never wholly shake off the religion of one's childhood; indeed there is little need when this is benign and still meaningful. Sadly this is far from universal.

Margaret Arden describes the work of Dr Montague Ullman, who founded a dream laboratory and later started the first of what is now a widespread collection of dream groups. I was impressed by her description of the way they work, which reminds me of my introduction to psychology in the seventies through attending a series of workshops in 'Transpersonal Psychology' run by Barbara Somers and others. Arden quotes from Ullman:

There are five underlying premises and three principles.

First premise. Dreams are intra-psychic communications that reveal in metaphorical form certain truths about the life of the dreamer, truths that can be made available to the dreamer awake.

Second premise. If we are fortunate enough to recall a dream we are then ready, at some level, to be confronted by the information in the dream. This is true regardless of whether or not we choose to do so.

Third. If the confrontation is allowed to occur in a proper manner the effect is one of healing. The dreamer comes into contact with a part of the self that has not been explicitly acknowledged before. There has been movement towards wholeness.

Fourth. Although the dream is a very private communication it requires a social context for its fullest realisation. That is not to say that helpful work cannot be done by an individual working alone, but rather that a supportive social context is a more powerful instrument for the type of healing that can occur through dream work.

Fifth. Dreams can and should be universally accessible. There are skills that can be identified, shared and developed in anyone with sufficient interest. Dream work can be effectively extended beyond the confines of the consulting room to the public at large.

It bears emphasising that dreams are intra-psychic communications. The process I use is geared to the needs of the dreamer as the one to whom the dream is being communicated. Communication to the group is a secondary affair necessary only for the group to make its contribution to clarifying the dream. So three principles can be stated:

1. respect for the privacy of the dreamer. Each stage of the process is designed to be non-intrusive so that the group follows rather than leads the

dreamer. The dreamer controls the process and there is no pressure to go beyond the level of self-disclosure which feels comfortable.
2. respect for the authority of the dreamer over his or her dream.
3. respect for the uniqueness of the individual.[7]

I would wholly endorse his words, particularly the need for dreams to be more widely accessed, and the comment made later that the current significance of dream symbolism should not be sacrificed to more general symbol associations. Not many, sadly, have the privilege of sharing their dreams in a knowledgeable and empathetic group.

Margaret Arden concludes an illuminating excursion into Goethe's thinking with the following profound passage:

> Consciousness can be seen as a sense organ for the perception of inner mental activity. The unconscious mind is therefore similar in kind to all the other natural processes of which we have knowledge. The distinction between mind and body is a distinction between different perceptual modalities. The brain does not exist separately from being experienced. The essence of human thought is the interconnectedness of things once the duality of alienation has been overcome through self-knowledge.
> If there is a unifying self, surely it is that self which creates our dreams … The only sensible view is that waking and dreaming consciousness are complementary forms of human activity … Paying attention to our dreams is one way of reversing the splitting processes that fragment society.[8]

Paying proper attention to dreams necessitates some defining creative action, which serves to anchor them in place. Jung spent a lot of his later life painting and sculpting his dreams in his retreat at Bollingen. I once spent five years rebuilding a stone wall, an activity I can recommend as a meditative exercise! As Goethe wrote, he did not feel that he had dealt with an experience until he had discharged it in creative activity.[9]

Myths can be regarded as collective dreams arising, according to Joseph Campbell, following the recognition of death:

> For it is simply a fact – as I believe we have all now got to concede – that mythologies and their deities are productions and projections of the psyche. What gods are there, what gods have there ever been, that were not from man's imagination? We know their histories: we know by what stages they developed. Not only Freud and Jung, but all serious students of psychology and of comparative religions today, have recognised and hold

that the forms of myth and the figures of myth are of the nature essential of the dream.[10]

The understanding and treatment of myths gave him much concern:

> The old differences separating one system from another now are becoming less and less important, less and less easy to define. And what, on the contrary is becoming more and more important is that we should learn to see *through* all the differences to the common themes that have been there all the while, that came into being with the first emergence of ancestral man from the animal levels of existence and are with us still.[11]

Campbell was writing in the 1970s. In the first decade of the new millennium it seems his message has not been read. Far from 'seeing through the differences', there is fragmentation of an obsessional, tribal kind taking place, and the old differences appear more and more important.

I have already given some of Joseph Campbell's quotation from the monk Thomas Merton. It is worthwhile here to give its antecedent:

> The symbols of the higher religions may at first sight seem to have little in common [wrote a Roman Catholic monk, the late Father Thomas Merton, in a brief but perspicacious article entitled 'Symbolism: Communication or Communion?']. But when one comes to a better understanding of those religions, and when one sees that the experiences which are the fulfilment of religious belief and practice are most clearly expressed in symbols, one may come to recognise that often the symbols of different religions may have more in common than have the abstractly formulated official doctrines.

In *The View from the Centre of the Universe*, Joel Primack and Nancy Ellen Abrams give a comprehensive review of the historical development of creation myths, or cosmologies as they prefer to call them.[12] They develop a new myth which they assert provides a realistic picture or parable of the history of the cosmos and our position and significance, as humans, within it. In doing so they hope to fulfil Joseph Campbell's dream of a unifying myth:

> In his last book, *The Inner Reaches of Outer Space*, mythologist Joseph Campbell made the passionate argument that what our society most desperately needs is a new story of reality for all of us – not just some chosen group. The story must demonstrate humanity's connection to all there is, yet be consistent with all we know scientifically. What he was longing for was a

new myth, but he knew that no one can simply create a myth, any more than they can 'predict tonight's dream'. A myth, he said, must develop from the life of a community. He hoped inspiration for such a story might come from physics. When he died in 1986, the revolution in cosmology was just beginning.[13]

Since it is impossible to comprehend the dimensions involved, it focuses on scale, or orders of magnitude, within which the solar system and mankind together occupy a central position.

> Most of us have grown up thinking that there is no basis for our feeling central or even important to the cosmos. But with the new evidence it turns out that this perspective is nothing but a prejudice. There is no geographical centre to an expanding universe, but we are central in several unexpected ways that derive directly from physics and cosmology.[14]

A number of fascinating conclusions emerge. One is a restoration of the Ptolomeic attitude. When one extends the view from the solar system to the cosmos, there is no longer any centre. The only meaningful centre is mankind, from which the cosmos can be seen to extend in all directions. Another is that since the cosmos constitutes everything that exists – and existence itself is paradoxical when one moves from one order of magnitude to another – there is no room for a God outside it. Insofar as the concept remains meaningful, God is as much a part of the cosmos as we are, in a sense both creature and creator. What the authors call 'the Cartesian Bargain' – the tacit assumption that, if there is a spiritual realm, it is separate from the physical universe – has been destroyed.[15]

When it comes to the origin of the universe one enters a realm of increasingly metaphorical language. Before the Big Bang, according to present hypothesis, there is postulated a zone of 'eternal inflation' where everything is in a potential state. The authors mention two possible 'myths' which follow as far as possible the scientific facts: the Cosmic Las Vegas and the Kabbalistic creation myth. Both require an exit from the potential state to an active state in which expansion in time suddenly becomes possible and materialises. The first proposes a random quantum possibility. The second, as far as I can understand it, is deterministic to the extent that it proposes the birth of a spark of (male) creative potential, *Keter*, within the eternal state (nothing but God), giving rise to a (male) flash of insight, *Hokhma*, 'the first thing separate from God'.

When it 'flashes into existence' there is created the womb of *Binah* which 'transforms the blueprint into the universe'.[16]

The concept of God itself, according to these authors, is renewed:

> The new scientific picture of the universe establishes a lower limit for God. For us, God can't be less or simpler, but could be more. God represents a maximum that is ever-expanding, and we are on the *inside*. God represents the *directions* of our wonder – not the destination … [17]

The idea that views of God expand with increasing scientific knowledge is nothing new in itself since all religions have been forced to come to terms with the expansion of scientific knowledge. However, the insistence that 'we are on the inside' is reminiscent of Jung:

> Now I knew … that man is indispensable for the completion of creation; that, in fact, he himself is the second creator of the world, who alone has given to the world its objective existence.[18]

However the authors continue:

> We two believe in God as nothing less than the process of opening our personal lines of contact with the unknown potential of the universe … We have a deep faith that if humans could come into harmony with the real universe, our troubled species would have its best chance to enjoy this jewel of a planet, unique in all the cosmos.

While I can follow them most of the way along the first sentence, the second has to overcome the obstacle of individual self-awareness: individuation, and I must hold with Jung that:

> If the whole is to change, the individual must change himself.[19]

Otherwise, as Jung warned, 'the artillery [or *Gaia*] will have the last word'.

One of the most interesting ideas expressed by Joel Primack and Nancy Ellen Abrams is the relationship between transcendence and a shift of scale, somewhat akin to what I used to call, using an older scientific metaphor, a 'phase shift'. The phase shifts between steam, water and ice do arise from a consideration of scale. The laws governing single molecules become irrelevant on a scale involving millions of molecules. When a system becomes sufficiently complex, 'emergent properties' (such as entropy) are observed and new laws apply.

Physical laws that apply on one scale do not cease to be true at other scales: they merely cease to matter.[20]

Moving from one scale to a higher can therefore be seen as transcendence. One is again reminded of Jung, this time of his 'transcendent function'. When opposite attitudes cannot be reconciled but are held in the imagination, often with suffering, then a symbol arises from the unconscious. The unconscious does indeed hold all scales.

The authors point to 'scale confusion' arising in many different areas, within and outside science. The concept of 'existence', for example, only applies around the centre of the scale spectrum, what they call *Midgard*. The question 'Does God exist?' demonstrates a confusion of scale. Likewise the question 'Does an electron exist?' is problematic:

> On a small scale, do electrons exist? The electron is a very useful concept – a particle with specific properties that we can talk about – but there is no solid thing, only a 'probability cloud'. In other words, the probability of its being somewhere is what's real. It doesn't make sense to say electrons exist in the commonly understood meaning of the word. But it makes even less sense to say they don't exist, because when you flick the switch, electricity flows and lights turn on. The confusion is due to the fact that when speaking of elementary particles, we lack any intuitive sense of their strange state and 'existence' is at best a metaphor.

Objects from another scale need to be called by us into reality, through symbols and metaphors, as the symbolic electrons did for the Doctor in Salley Vickers' novel quoted earlier.[21]

On the social plane there is constant scale confusion. The kind of *entente* among soldiers in a company cannot be assumed to exist in a division. That has been recognised in the army from the time of the Roman *centuriae* and given its place in the traditional chain of command. Every soldier will at least know his colonel, will meet him on occasion and feel loyalty to his regiment. Every commander up the scale is imbued with responsibility and loyalty up and down the scale. Such a sense of loyalty and belonging is almost absent in democratic civilian life, at least since the demise of the squirearchy. Everyone knew the squire as the head of an extended family. Country villagers still, mostly, know the chair of the parish council; rather fewer in towns, though the appointment of a mayor of London was an important stroke. But the concept of a constituency or even a district council represented by individual voters seems to me to fall foul of scale confusion.

This seems to me compounded by a lack of appreciation that each level in the scale of authority requires for its validity a spiritual element represented by a leader having the necessary vision. Some are born great, as Malvolio ruminates, some acquire greatness, and some have greatness thrust upon them; but rarely is greatness acquired by personal ambition and self-advertisement. Exceptions, as always, prove the rule but self-interest and corruption in corporate and public office are almost a norm and are reflected by public apathy.

In a monarchy like the United Kingdom, the disassociated institutional fragments are held precariously together by the King or Queen. It is often forgotten that the army (which I am using in its ancient sense of including all the arms) owes allegiance not to the government of the day, the establishment, but directly to the monarch. It serves 'King and Country'.

The King is one of the oldest and greatest of all symbols. The risen Christ is venerated as King of Heaven. The King is not chosen by vote from among the populace, but takes his place by divine right of inheritance. He arrives prepared to rule with his loyalties and responsibilities ingrained since birth and is anointed by the High Priest in an ancient and traditional ritual with antecedents going back to David and Solomon, and beyond.

> Zadok the priest and Nathan the prophet anointed Solomon King.
> And all the people rejoiced and said:
> God save the King!
> Long live the King!
> May the King live forever![22]

It is doubtful whether the present monarchy could survive in the absence of an established Church.

Charles I of England abused his position but was nonetheless correct in seeing his appointment, at least in a symbolic sense, as divine. With the subsequent advent of constitutional monarchy, the role of King or Queen is all the more symbolic, but not less significant. Perhaps more so. It adds the spiritual quality necessary to bind the souls of the 'subjects' into nationhood and endows the concept of Nation with a mythic quality. The King, says Cirlot:

expresses the ruling or governing principle, supreme consciousness, and the virtues of sound judgement and self-control. At the same time, a coronation is equivalent to achievement, victory and consummation. Hence any man may properly be called a king when he achieves the culminating point in the unfolding of his individual life ...

Love also plays a highly important part in the symbolism of royalty, since love is held to be one of the most obvious of culminating points in the life of Man.[23]

'The further we go back in history,' says Jung,

the more evident does the king's divinity become. The theology of kingship best known to us, and probably the most richly developed, is that of ancient Egypt, and it is these conceptions which, handed down by the Greeks, have permeated the spiritual history of the West. Pharaoh was an incarnation of God and a son of God. In him dwelt the divine life-force and procreative power, the *ka:* God reproduced himself in a human mother of God and was born from her as a God-man. As such he guaranteed the growth and prosperity of the land and the people, also taking it upon himself to be killed when his time was fulfilled, that is to say when his procreative power was exhausted.[24]

The Nation is, after all, the inheritor of the tribe which dominated humanity for millennia. We are still but the gloss on so many thousands of years of tribalism. The mythic quality of the tribe is amply demonstrated by the failure of 'constructed' nation states like Iraq, the Soviet Union or the ill-fated East African Federation of the 1950s. Such states, split into traditional tribal units and, lacking any overriding spiritual unity or mythic tradition, can only be held together by a demonic autocrat or a figurehead of sufficient charismatic quality, a King in name or all but name. This quality has, over the years, been imparted to the position of President in certain more successful nations, notably the United States, where the necessary 'anointing' as dedication to the service of his people takes place at the oath of office. But youthful democracies are always precarious.

With the huge increase in migration taking place, not only in Europe but around the world, bringing with it competing self-interested bodies with their own tribal associations, the future for the nation state also looks precarious. I am not clear what will take its place. I fear that the presently existing global tribes are far from ready for a universal cosmological myth, with God 'within', attractive as I – and I am sure many others – may find it.

The present heir to the throne of the United Kingdom, Prince Charles, said some years ago that he would prefer to have the title 'Defender of Faiths' to the present royal title 'Defender of the Faith' – always uncomfortable since it was bestowed by a pope shortly before Henry VIII proclaimed himself head of the Church of England. It may yet be forced upon him. In the Reith lecture in 2000, at the start of the new millennium, he also said:

> The idea that there is a sacred trust between mankind and our Creator, under which we accept a duty of stewardship for the earth, has been an important feature of most religious and spiritual thought throughout the ages. Even those whose beliefs have not included the existence of a Creator have, nevertheless, adopted a similar position on moral and ethical grounds. It is only recently that this guiding principle has become smothered by almost impenetrable layers of scientific rationalism. I believe that if we are to achieve genuinely sustainable development we will first have to rediscover, or re-acknowledge a sense of the sacred in our dealings with the natural world, and with each other. If literally nothing is held sacred anymore – because it is considered synonymous with superstition or in some other way 'irrational' – what is there to prevent us treating our entire world as some 'great laboratory of life' with potentially disastrous long term consequences?[25]

The associations of the diamond and star with royalty have been illustrated in Chapter 3. Prince Charles has remained faithful to his affirmed sacred trust against much criticism and even vilification. But I fear that a 'Defender of Faiths' will perpetuate the current fragmentation. What is needed is, of course, a unifying myth. But to reach it will require more individual self-awareness, which I believe needs as a first step a symbolic attitude.

NOTES

1 Margaret Arden, *Midwifery of the Soul*, Free Association Books Ltd., 113.
2 Peter J. Conradi, *Iris Murdoch: A Life*, HarperCollins.
3 Fellow of All Souls and author of *Journey into Ladakh*.
4 Conradi, *ibid.*, 552.
5 *Ibid.*, 554.
6 *Ibid.*, 554-5.
7 Arden, *ibid.*, 93, from Ullman, M. [1973], 'Dreams, the Dreamer and Society',in *New Directions in Dream Interpretation*, State University of New York Press.
8 Arden, *ibid.*, 113.
9 *Ibid.*, 108.
10 Joseph Campbell, *Myths to Live By*, Viking, 253.
11 *Ibid.*, 24.
12 Joel Primack and Nancy Ellen Abrams, *The View from the Centre of the Universe*, Fourth Estate, Chapter 2.
13 *Ibid.*, 33.
14 *Ibid.*, 7.
15 *Ibid.*, 79.
16 *Ibid.*, 200.
17 *Ibid.*, 277.
18 C.G. Jung, *Memories, Dreams, Reflections*, RKP, 240.
19 See Introduction.
20 *Ibid.*, 167.
21 See Chapter 5.
22 From Handel's oratorio, based on I Kings, 1, and sung at the coronation service.
23 J.E. Cirlot, *A Dictionary of Symbols*, RKP, 167.
24 C.G. Jung.
25 Prince Charles, Reith Lecture, 2000.

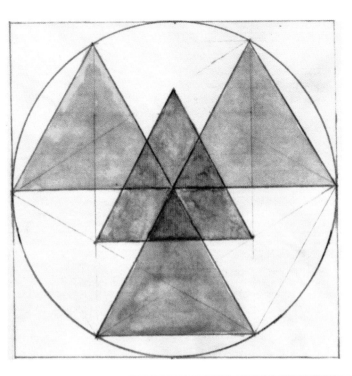

PLATE 1

The diamond
and star
mandalas

Diamond
mandala
John Warden,
1988

Star mandala
John Warden, 1988

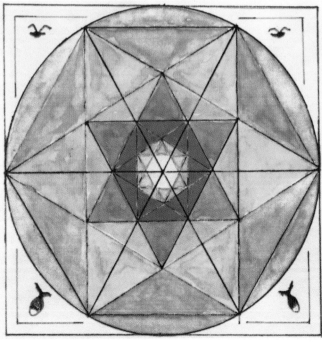

PLATE 2a

Mandala

Reproduced from
*The Secret of the
Golden Flower*

PLATE 2b

Tibetan mandala
Reproduced from
www.dharmanet.com.br

PLATE 2c

Hawaiian mandala
Photograph by John Warden

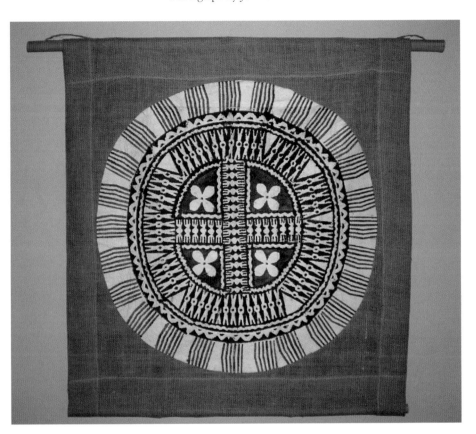

PLATE 2d

Tracing from electron micrograph
of a chick liver cell showing
the membrane and nucleus
Reproduced from
Steven Rose, *Lifelines*

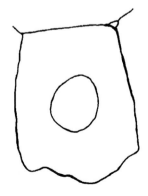

PLATE 3a

Prima materia
Reproduced from
de Rola, *Alchemy*, Flare

PLATE 3b

Conjunctio or
sacred marriage
Reproduced from
de Rola, *Alchemy*, Flare

PLATE 4a

Uroboros
Reproduced from
www.azothgallery.com

PLATE 4b

The New Uroboros
Reproduced from Primack and Abrams, *The View from the Centre of the Universe*

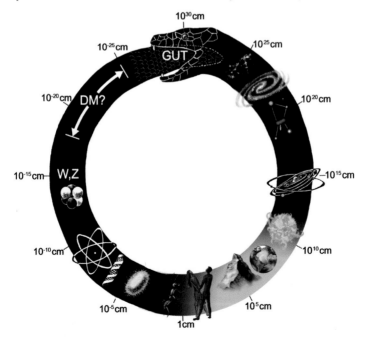

PLATE 5

Danseuses by Matisse. Reproduced from www.perso.wanadoo.fr

PLATE 6

Free energy landscape of folding of protein segment, 'the tryptophan zipper'. Stable, low-energy 'valleys' shown in blue.

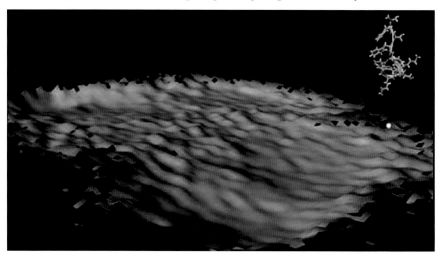

PLATE 7

Ribbon diagrams of the enzyme phosphorilase in two folded states
from the crystal structures

Reproduced from Geoffrey Zubay, *Biochemistry*,
3rd edn, William C. Bown Communications Inc, Iowa, USA

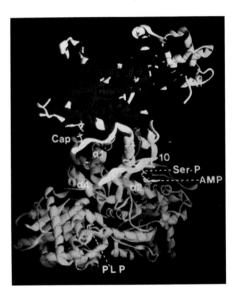
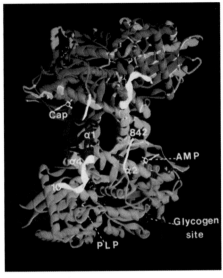

PLATE 8

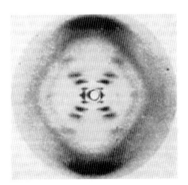

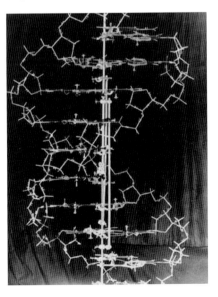

Rosalind Franklin's diffraction
photograph of the DNA crystal
structure (B structure)

Watson and Crick's model of
the DNA sructure

Photographs: Cold Spring Harbor
Laboratory Archives

PLATE 9

Diagram of the I Ching

Central image reproduced from www.rotten.com

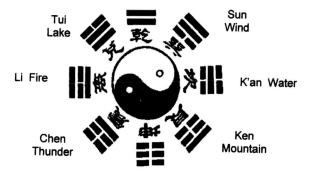

Ch'ien Heaven

Tui Lake

Sun Wind

Li Fire

K'an Water

Chen Thunder

Ken Mountain

K'un Earth

PLATE 10

Dream image – The Holy City

Painted by John Warden

PLATE 11

Dream image –
'Alpha'
Painted by
John Warden

PLATE 12

Carbon-carbon bonds

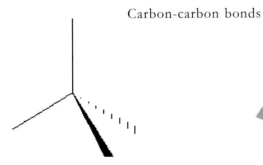

The four valencies

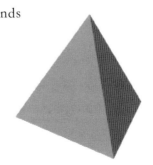

Tetrahedron

PLATE 13

Carbon-carbon bonding
in the diamond

Crystalline graphite

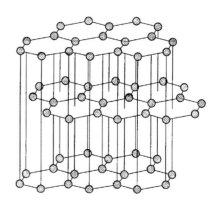

PLATE 14

The phases of carbon

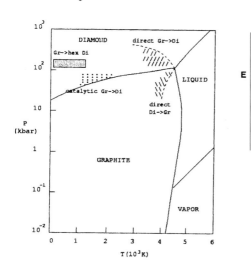

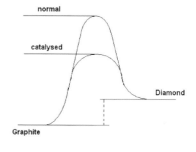

Energy diagram of
the two crystalline
phases of carbon

PLATE 15

Carbon compounds

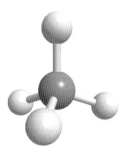

Methane

Benzene   Cyclohexane   Cyclohexane
               'chair'   'boat'

CH₂OH   CH₂OH
Glucose   Fructose

Sucrose

PLATE 16

Starch and Glycogen (α-1, 4 linkages)

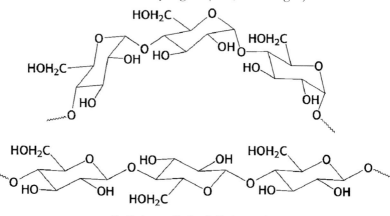

Cellulose (β-1, 4 linkages)

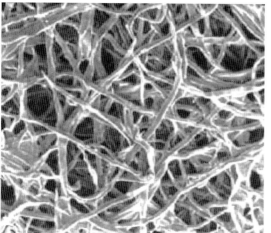

Cellulose

PLATE 17

Buckmaster Fullerite

PLATE 18a

Mercury as the
evening star
Reproduced from
www.astronomy.com

PLATE 18b

Venus as the
evening star
Photograph:
John Warden

PLATE 19

Dream image – Initiation
Painted by John Warden

PLATE 20a

Temple Gate, Japan
Photograph: John Warden

PLATE 20b

Spinning Tai Ji
Reproduced from Chungliang Al Huang,
Quantum Soup

Chinese character *men* (gate)
Reproduced from Chungliang Al Huang,
Quantum Soup

PLATE 21

Angel's wing
Photograph: John Warden

8
Human Differences

I FIND IT DIFFICULT to conceive an attitude that fails to recognise that human beings differ psychologically at a profound, even genetic level in such a way that one can recognise different types into which one can classify them. Yet it would seem that this is not universally accepted. Perhaps I am writing from a point of view somewhat ingrained in Jungian thought. However it struck me, reading in Margaret Arden's book *Midwifery of the Soul* about Gregory Bateson's and Matte Blanco's categories of thought, that, although their ideas were persuasive, I would have reconstructed their thinking in terms of typology.

Bateson seems to have been the author of the 'double bind', the concept of confusing or contradictory messages so commonly experienced in childhood which can lead to trauma or even madness.[1] This surprises me a little since it seems similar to much of R.G. Laing's earlier writing on schizophrenia and indeed Winnicotts's writings on the 'false self', which also seem earlier.[2] Bateson followed Russell's theory of logical types to distinguish signals exchanged between people. Thus words spoken at an explicit level may be contradicted by gestures, tone or context at an implicit level. As I understand Matte Blanco's work, he distinguished between asymmetric thinking, which uses logic, and symmetric thinking, where logical processes do not apply.

> The concept of symmetry is particularly relevant to affects, and Matte Blanco describes experiences of strong emotion as infinite experiences. It is well known that under stress of strong emotion logical thinking becomes very difficult ...
>
> Asymmetrical and symmetrical thinking co-exist in all mental activity in varying proportions ... A lover does not see any limits to his love; it is unbounded, and the fact of his love colours his whole existence. At the same time he has asymmetrical knowledge of the fact that he is in love.[3]

While these are useful paradigms with which to discuss conscious and unconscious psychic activity, it seems to me that they suffer from the drawback that they treat such activity as the same for all, something that Jung was insistent was mistaken or inadequate early in his career. His book *Psychological Types* was published in 1921, not so long after his split with Freud. Indeed this was based on a paper delivered at the congress in Munich in 1913, the last time they met.[4]

He first distinguished the differing attitudes of introversion and extroversion, from his observations of schizophrenia and hysteria in his psychopathic patients at the Burghölzli Institute. (Unlike Freud, who mainly treated neurotics, Jung's early experience was in psychopathology.) According to C.T. Frey, one of the best Zürich analysts, now deceased, Jung said it was because he saw the inevitable break between Freud and Adler approaching and wanted to offer a viewpoint from which both standpoints, both theories, were acceptable:

> He wanted to avoid a game of shadow projections between the Freudians and Adlerians, and you can see what happened: he got into deeper trouble himself![5]

Adler was of course the extravert and Freud the introvert. In Jung's original words:

> To this extent, extraversion and introversion are two modes of psychic reaction which can be observed in the same individual. The fact, however, that two such contrary disturbances as hysteria and schizophrenia are characterised by the predominance of the mechanism of extraversion and introversion suggests that there may be also normal human types who are distinguished by the predominance of the mechanism of one or other of the two mechanisms.[6]

In his book (CW 6) Jung goes on to compare real and mythical figures. Plato, for whom only ideas were real, he saw as introverted, and Aristotle, who took outer things as real, he saw as extraverted. Schiller was an introvert, Goethe an extravert. Following on from the two attitudes, he went on to examine different ways of viewing an object and defined the four functions. Two modes of perceiving he called irrational: sensation and intuition; the other functions which deal with evaluation or apperception he called the rational functions: thinking and feeling.

Sensation, perceiving, e.g. by touch or vision, needs no explanation; nor does thinking, evaluation through cause and effect. The other two are not so simple to grasp. Jung's use of the expression 'feeling' (I am not sure of the original German) needs to be differentiated from emotion. It is a matter of judgement, but a holistic judgement, short circuiting cause and effect. In fact the Myers-Briggs type test substitutes judgement for feeling. Likewise intuition is difficult. Like feeling it is a holistic form of perception: seeing the wood rather than the trees.

This classification is not without its critics. C.T. Frey remarked in the same lecture:

> If you feel warm for another person, that may have to do with the fact that you like this person, but I still think that this warmth is something other than the evaluating 'I like it' or 'I don't like it', which Jung means here.
>
> And after that, we have to decide where that thing comes from or where it goes, and for this we need intuition. Now, intuition is the greatest problem of all. Jung says that all these four functions are simple psychological concepts. Then he goes on and says intuition is not thinking, it is not sensing and it is not feeling – even though it is of course all the three of them. Now if you call that a single function, I don't know what a single function is. That's the problem with the types.
>
> Again, I do not mean that the types don't make sense, but they're not yet worked out and not yet clear.

Intuition is probably the most difficult of the psychological functions to pin down intellectually, in spite of being quite widely accepted as an experience, not least by scientists. I have already mentioned Henri Poincaré and Kekule but it is worth citing other remarks of theirs:

> Often nothing good is accomplished at the first attack. One takes a rest; and then all of a sudden the decisive idea presents itself to the mind.
>
> Henri Poincaré

> Let us learn to dream, gentlemen; then we shall perhaps find the truth.
>
> Friedrich Kekule

According to Mary Midgley, Kekule saw a serpent eating its tail. This is the Uroboros, a very archaic symbol which Jung regarded as one of the basic symbols of alchemy,[7] sometimes taken as portraying a sudden goading from the unconscious. The story I prefer is that the vision was of dancers in a ring like the famous painting by Matisse,[8] which mimics the bonding of the atoms and the circular flow of energy (Plate 5).

Intuitions from the unconscious are not always recognisably symbolic; they may arrive as an idea rather than an image. But they all require exercise of the imagination, followed by some hard thinking to bring them to a full realisation. If this amounts to a hypothesis, then the experimentation begins. The second quotation from Poincaré is reminiscent of the Zen experience of being 'stuck', which is considered to be a blessing. The linear process of thought can get no further. One has to stop; do something quite different, and suddenly the answer emerges. I often go to bed in some depression with all my ideas in confusion. Yet as soon as I get down to work in the morning, everything is suddenly clear. I imagine this as the work of my little ants, sorting things out for me, as they did for Psyche.[9] There is so much of value in the old myths, disregarded as they are, and at such peril.

Joel Primack and Nancy Ellen Abrams, in their book *The View from the Centre of the Universe*, have some interesting thoughts on intuition. Their view is that, together with common sense, intuition needs continually to be brought up to date. There is certainly a danger of remaining obstinately within an old paradigm, like flat-earthers:

> ... our intuition reclines in an old, overstuffed, and unaffected image of reality. Everyone knows, for example, that the earth goes round the sun, but we still talk about sunrise and sunset. Language is powerful and it can reinforce obsolete intuitions, as those words do – or it can challenge them.[10]

I am not entirely persuaded. I would not regard sunrise and sunset as intuitions, merely poetic descriptions of natural observations. The sun *appears* to rise and set. It took intuition for Copernicus to turn this idea inside out.

The authors identify intuition with a desire to fit ideas, or select theories, into whatever seems most beautiful. There is certainly force in this. They plead for an attitude which they call 'counterintuitive', which seems the same as that I am advocating as an 'open mind', which, by comparison, seems a boring, old-fashioned idea and inadequate for what I have in mind. That is why I have extended my plea to acquiring a symbolic attitude, an attitude like that of the Boy Scout of being constantly prepared for the unexpected, for the intrusion of wonder. Their meaning of counterintuitive allocates to intuition an attitude of *jumping to conclusions*, based on common sense, which I agree needs constant revision. But this is closer to a form of judgement or feeling in a Jungian sense, not intuition as I seek to explain it – in fact almost its opposite!

This is a quarrel over semantics, yet how important they can be. The importance, as I see it, is that, by defining intuition downwardly towards judgement, they leave no room for the truly intuitive, the 'bolt from the blue', the sudden symbolic image, thought or feeling. As will become more apparent, intuition has a number of meanings, even within Jung's writings.

Perhaps the poets best express the feeling function – or does this result, as Frey has indicated, from confusion between judgement and emotion? How does one distinguish between 'I like this' and 'I don't like this' without involving some emotion, some upthrust of energy from the unconscious? By using the word 'feeling' I believe Jung must have accepted this. He could have used 'judgement' or some expression such as 'intuitive discrimination'. Again one can't pin it down intellectually. Perhaps one shouldn't even try. Poetry, like music, is essential to human wellbeing, 'the food of love'.

Mary Midgley, in *Science and Poetry*, used the poet as midwife in another sense – between our inner, personal world and the larger public world outside us, and between thought and feeling, which gave rise to many 'false antitheses' during the Age of Reason:

> All the great Romantics made this effort to bring both sides together, which is just what makes them great. Wordsworth and Coleridge in particular went to great lengths to stress that the antithesis between thought and feeling was a false one. They insisted that both were aspects of a single whole that might best be understood by attending closely to its middle term, imagination. Here was the scene of the process of creation, both in art and science – not a mass of idle and delusive fancy, but a constructive faculty, building experience into visions which made both feeling and thought effective. A poet, said Wordsworth, had to be 'a man who, being possessed of more than usual organic sensibility, had also thought long and deeply … Our thoughts … are indeed the representatives of all our past feelings'.

In short, there is no necessary battle between our different parts; there is merely a great difficulty in seeing ourselves as a whole. Wordsworth therefore insisted that poetry needs a skeleton of serious thought.[11]

Here is a great 'thinker' summoning up her inferior function and doing battle!

Poetry seems to validate the necessary human function of feeling, permitting the entry of imagination and emotion to round out a holistic view of any subject or object 'in mind'.

Grasping the differences among the types and attitudes is not easy. There are eight possibilities: introverted sensation, extraverted sensation, introverted and extraverted feeling, and so on, and these differ quite considerably so that Jung takes a great deal of space to sort them out. And that is not the end of the matter, since he ascribes to each individual four function-priorities: the main or principle function, the auxiliary function, and then the less-used third and fourth functions, the so-called 'inferior functions'. The latter two are substantially unconscious and come into operation when the unconscious takes over, as in affects, or in psychopathology, which Jung used as a mirror. As one gets older, past the midlife change, one's functions alter as the older ones fall into disuse. Also, as one individuates, one becomes, hopefully, less of a 'type' as one brings the unconscious functions into greater use.

Various type tests have been devised in which a series of questions is asked. The first was probably that produced by Horace Gray and Joseph Wheelwright. This has been overtaken in use by the Myers-Briggs test. Frey criticises the former as not being very good on the thinking-feeling scale and generally prefers the latter. But no such tests are infallible and it must be remembered that types are just that: no individual is an exact type. As Frey goes on to say:

> I think the whole type business is extremely controversial and shaky. But it is useful to think in these terms – to use them, to play with them. It is less useful to pin yourself or somebody else down.

It is nevertheless amusing to 'type' ones friends and acquaintances and many funny stories have been written around the functions. One from Jung can be found at CW 7, paragraph 81. A more amusing one is given by Henri Ellenberger in his great work, *The Discovery of the Unconscious*:

> As a mnemotechnic device Ania Teillard imagined the story of the dinner of

the psychological types: The perfect hostess (feeling-extroverted) receives the guests with her husband, a quiet gentlemen who is an art collector and expert in ancient paintings (sensation-introvert). The first guest to arrive is a talented lawyer (thinking-extroverted). Then comes a noted businessman sensation-extroverted) with his wife, a taciturn, somewhat enigmatic musician (feeling-introverted). They are followed by an eminent scholar (thinking-introverted) who came without his wife, a former cook (feeling-extrovert), and a distinguished engineer (intuitive-extroverted). One vainly waits for the last guest (intuitive-introverted) – but the poor fellow has forgotten the invitation;[12]

Ellenberger mentions that 'basic to Jung's conception was his personal, real life experience of the process of increasing introversion and the return to extraversion in the course of his creative illness'. Looking back over my life, I can recognise fluctuations in attitude. As a young child I was happily extravert, but the trauma of an early separation to a boarding school induced an introvertive shift which stayed until my confidence grew, when I became increasingly extraverted through my prime. Then, after a 'mid-life crisis', I became quite introverted again, though in a more balanced way.

Likewise, while (like a good Taurus) my main function has always been sensation and I have thought of myself as a slow thinker, I realise that in the pursuit of my primary occupation (as a patent attorney) I had to work really very fast in an analytic way. I also had to take an incisive overall view of a concept presented to me usually in a multitude of details. In all of this I relied heavily on my unconscious – my intuition – but must also have done a lot of quick thinking! Now I am quite content to think slowly again, though my reliance on intuition and feeling have increased hugely.

Ellenberger also points out that others had experimented with typology, mentioning Janet, Bleuler, Kretschmer and Rorschach, describing a fascinating experiment by Binet on his young daughters. Binet (1903) used the words 'introspection' and 'externospection'. I do not know what views Freud had on typology and cannot find any reference to these in Ellenberger. Neither can I find any mention in Margaret Arden's book (above). It seems to be a subject foreign to (Freudian) psychoanalysis.

What I find fascinating is the close parallel with the four principle signs or 'humours' of astrology and the sun signs:

SUN	SIGN	TYPE
Aries	Fire	Intuition
Taurus	Earth	Sensation
Gemini	Air	Thinking
Cancer	Water	Feeling
Leo	Fire	Intuition
Virgo	Earth	Sensation
Libra	Air	Thinking
Scorpio	Water	Feeling
Sagittarius	Fire	Intuition
Capricorn	Earth	Sensation
Aquarius	Air	Thinking
Pisces	Water	Feeling

There are further parallels with alchemy and with the four humours of old medicine. Jung may well have had all these in mind when working out his types. When considering the possible type of a friend I always try to discover his sun sign as a first step. It is very illuminating, if sometimes quite misleading. Astrology has assumed a popularity with the advent of the New Age after being for many years anathematised by both science and religion. Jung pays considerable respect to astrology, which he saw as an effect of synchronicity. In his memorial address to Richard Wilhelm he remarks that '… astrology represents the summation of all the psychological knowledge of antiquity':

> The fact that it is possible to reconstruct in adequate fashion a person's character from the data of his nativity shows the relative value of astrology … Insofar as there are any really correct astrological diagnoses, they are not due to the effects of the constellations but to our hypothetical time qualities. In other words, whatever is born or done in this moment of time has the quality of this moment of time.[13]

While I remain sceptical of the minutiae of astrological prediction, it has always seemed to me quite plain that people fall into certain categories of personality, and that these should reflect the quality of a particular moment in the earth/sun system, or in the universe, seems intuitively right (in Primack and Abrams' sense of beautiful) or feels right (in Jung's sense). It may, of course, be misleading, but that awaits the next dimension of understanding.

The great problem, I believe, in Jung's analysis of psychological types is the use of the same words for quite different phenomena. The worst is perhaps 'intuition'. The types and the functions they appropriate portray how human personalities differ in their basic outlook on the world – their *weltanschauung*. An 'intuitive type' describes a person who (ideally) views the world in a certain way which differs from that of a 'sensation type', i.e. he tends to see objects as ideas or concepts; his vision is broad and he ignores details where they interfere with this vision. I have given some illustrations, but here is one more. Perhaps an intuitive approaching the ideal was that great intuitive, E.F. Schumacher (Leo). He could see the whole effect of an ever-expanding technology on human life and the human spirit:

> The UN and the World Bank produce indices of urbanisation, showing the percentages of the population of different countries living in urban areas (above a certain size). The interesting point [*for Schumacher*] is that these indices entirely miss the interesting point. Not the degree but the *pattern* is the crux of the matter [bracketed italics mine].[14]

Schumacher conceived the idea of intermediate technology, but left the details to his sensation friends. The difference from Marx, whom he quotes with some approval, is that this Taurus met the devil in the detail.

But this use of 'intuition' seems to me to have little to do with the kind of intuition, an upsurge from the unconscious, that affected Kekule and Henri Poincaré – both astrologically sensation types. Indeed (as I know from experience) the kind of 'inferior' intuition that can reach a sensation type on a surge of emotion, tainted though it may be with the dark of the shadow, and provided it is recognised and not immediately projected, is like gold dust.

In a similar way, the gift that makes a great poet, his human warmth, should not be confused with his basic outlook. Poets differ hugely in style: the great introverted thinker T.S. Eliot (Libra) is so completely different from those masters of verbal sonority, the extraverted sensation types Thomas Gray and Rudyard Kipling (both Capricorn and my favourites).[15] Once this confusion is rectified, the types make more sense.

To return now to Margaret Arden, one of the properties of the double bind, seen in Jung's terms, is the dichotomy between intuition and thinking. The mother gives an order explained in terms of cause and effect, but the child experiences a contrary order through its intuition of the mother's unconscious attitude. However I find it difficult to believe that Jung's types have much relevance in childhood, where the child lives so much in his or her unconscious, and cause and effect are at a very early learning stage. This (so it seems to me) shows up the difference in focus between Freudian and Jungian psychology. Jung was much more concerned with adult psychology and his types are only really appropriate once development has reached a point where the personality has to a large extent crystallised. This, of course, is not to devalue them, especially in a general discussion.

Arden goes on to discuss David Bohm's theory of implicate order:

[In nuclear physics] The observer and the event observed are part of a larger situation, in which objective information about the thing observed is not possible. It is only possible to describe what was observed under certain conditions. Knowledge too is a process, an abstraction from the total flux which is the ground both of reality and of the knowledge of reality. Bohm uses the word 'holomovement' to describe the totality of our idea of what is. By definition we can only have fragmentary knowledge of the holomovement.

Bohm uses the metaphor of a patterned carpet in which flowers and trees are represented. The relevant way to look at the carpet is to be aware of the pattern, and it is not useful to say that the various motifs are separate objects in interaction.[16]

Bohm's concept of holomovement, as far as I understand it, is related to Primack and Abrams' scale confusion. Looking at the details and looking at the pattern are forms of observation on differing scales. They are both perfectly valid, but require quite different understanding, under different laws or frames of reference.[17]

The patterned carpet is like the wood full of trees I mentioned earlier. Jung's intuitive would receive an immediate impression of the pattern, whereas a sensation type might well be fascinated by the details. Surely only an intuitive could write '*it is not useful*' to say that the various motifs are separate objects in interaction'! (It is not clear to me whether the words are Arden's or Bohm's.)

The metaphor of the carpet reminds me of a parable (I am not sure where I heard or read it). A caterpillar was creeping slowly across the ground, noticing how the colours changed: now it was a cheerful red; now a gloomy black. He thought it most peculiar. Then one day he rose in the air as a butterfly and lo: there beneath him he saw his life laid out as a beautiful pattern.

Readers' reactions to this book and to symbols of the diamond and the star will undoubtedly vary with their typology. All an author can hope is that all types can find something that rings true.

NOTES

1 Margaret Arden, *Midwifery of the Soul,* Free Association Books Ltd., 10.
2 *Ibid.,* 110.
3 *Ibid.,* 12, 13.
4 The original lecture is published in English in CW 6, 499.
5 C.T. Frey, lectures at the Jung Institute, Zürich (unpublished).
6 C.G. Jung, CW 6, 862.
7 Plate 4a.
8 Plate 5. (Matisse himself was not born until after Kekule's vision, or dream as some describe it.)
9 See Chapter 7.
10 Joel Primack and Nancy Ellen Abrams, *The View from the Centre of the Universe,* Fourth. Estate, 32.
11 Routledge edition, 55.
12 Henri F. Ellenberger, *The Discovery of the Unconscious,* Harper Torchbooks, 702.
13 Richard Wilhelm, *The Secret of the Golden Flower,* RKP, 142.
14 E.F. Schumacher, *Good Work,* Abacus, 46.
15 Out of interest I have dated and 'typed' astrologically some other poets but with less clarity:

           Intuitives   Shakespeare, Tennyson, Wordsworth, Shelley
           Thinkers   Masefield,
           Feelers   Keats

   Intuitives seem to predominate. I would have classed Masefield as sensation.
16 Arden, *ibid.,* 29.
17 Discussed in more detail in Chapter 7.

9

Two-Dimensional Symbolism:
The Star

Not where the wheeling systems darken,
And our benumbed conceiving soars –
The drift of pinions, would we hearken,
Beats at our own clay-shuttered doors.[1]

W HILE ATTENDING some lectures on mandalas in the 1980s
I was struck suddenly by the planar, or two-dimensional,
emphasis throughout the symbology of the mandalas illustrated and discussed. Jung gives pride of place to the mandala as a symbol
of the Self. Although there is no complete agreement on essentials of a
mandala (Plates 2a-c),[2] it is accepted that it involves the imposition of
a quaternary structure on the circle; According to Cirlot,[3] the circle
represents the universe, the All (including, of course, all that is conscious
and unconscious); the quaternary generally represents the plane of the
earth, the material pattern of life.

But essential to the mandala is the spiritual centre, the nucleus, representing the Godhead, to which the gates in the periphery open and all
the roads lead:

> In all symbols expressive of the mystic Centre, the intention is to reveal
> to Man the meaning of the primordial 'paradisal state' and to teach him to
> identify himself as the supreme principle of the universe. This centre is in
> effect Aristotle's 'unmoved mover' and Dante's 'L'Amore che muove il sole
> a l'altre stelle'. Similarly Hindu doctrine declares that God resides in the
> centre, at that point where the radii of a wheel meet at its axis. In diagrams
> of the cosmos, the central space is always reserved for the Creator.[4]

In the mandala from The Secret of the Golden Flower (Plate 2a), the centre is represented by the *dorje*, the diamond thunderbolt, that strange symbol which seems to indicate the explosive reaction of pure energy and pure matter in the act of creation. But the spiritual centre is the star.

> As a light shining in the darkness, the star is the symbol of the spirit … Jung recalls the Mithraic saying 'I am a star which goes with thee and shines out of the depths'.[5]

The star can therefore be seen as symbolising the spiritual nucleus or creative force at the heart of the two-dimensional mandala, the Self symbol of the Hindu and Budhist religions pervading the East.

Jung's fascination with the mandala was closely associated with the importance he attached to the concept of the quaternary, which he viewed as symbolising twin polarities, the concept that every property has, or gives rise to, its opposite; that in the dynamic tension between the opposites lies the human psychological drama. This has been explored already in Chapter 3. The concept is present also at the heart of Taoism, although 'dynamic tension' is a Jungian or Western perception. Taoism perceives the relationship differently – more like a dance:

> When all men recognise beauty, ugliness is conceived.
> When all men recognise goodness, evil is conceived.
> So existence supposes non-existence;
> The difficult is the complement of the easy;
> The short is the relative of the long;
> The high declines towards the low;
> Note and sound relate through resonance;
> Before and after relate through following;
> So the wise man needs no force to work,
> And needs no words to teach.
> *Tao Te Ching*, Chapter XI

Some months before writing my original essay I made a study of the *Tao Te Ching* of Lao Tsu, comparing different translations: Ch'u Ta-Kao (Allen & Unwin), D.C. Lau (Penguin) and Richard Wilhelm (Arkana). In some instances, quotes in Alan Watts' words are from *Tao: The Watercourse Way* (my preferred authority). However I found the divergence in the translations so great that it was sometimes difficult to recognise an identical source. This is inevitable, given the pictorial nature of Chinese calligraphy and indeed the humour of Lao Tsu, who constantly uses puns.

Both Taoism and Zen Buddhism give great attention to laughter and
Chungliang Al-Huang opens his splendid book *Quantum Soup* with a
discourse on laughter, which reminds me also of the great Western
philosopher and theologian Erasmus' book *In Praise of Folly*. No one
having read *Quantum Soup* could ever take himself too seriously again.

> *Laughter* in Chinese writing is depicted by a human with arms and legs flung
> wide apart. head up to the sky, vibrating with mirth like bamboo leaves in
> the wind.[6]

> One of the most highly recommended Taoist and Zen meditations is to let
> your hair down, stick out your belly and roar with laughter.[7]

The hermeneutic difficulties involved have been explored by J.J. Clarke
in his book *Jung and Eastern Thought*. He quotes from Gadamer:

> To try to eliminate one's own concepts in interpretation is not only impos-
> sible, but manifestly absurd. To interpret means precisely to use one's own
> preconceptions so that the meaning of the text can really be made to speak
> to us.[8]

So I took to rewriting the work in my own words, using Jungian insight,
since I recognised the close parallel and psychological foundation of the
Taoist writing. The quotation above is in my own words.

There seems to be a quality in Lao Tsu which invites the reader to
express the images in the most meaningful personal sense; reading it is
a personal experience. A vivid example of a highly personal experience
is given in Robert Pirsig's novel *Zen and the Art of Motorcycle Maintenance*.[9]

The origin of many of Jung's ideas lie in gnosticism, for which he later
found much support in alchemy, which he regarded as the successor of
gnosticism. However his ideas may have crystallised in that awesome
product of his active imagination, the *VII Sermones ad Mortuos* (published
rather late in 1925 and before he read Richard Wilhelm's *Secret of the
Golden Flower*), in which the pleroma is made manifest through the terrible god Abraxas:

> We must therefore distinguish the qualities of the pleroma. The qualities are
> PAIRS OF OPPOSITES, such as –
>> The Effective and the Ineffective
>> Fullness and Emptiness

> Living and dead
> Difference and Sameness
> Light and Darkness
> The Hot and the Cold
> Force and Matter
> Time and Space
> Good and Evil
> Beauty and Ugliness
> The One and the Many ...   Sermo I

Abraxas is effectiveness, more awesome in Jung's conception than the gentle way of the Tao, very much a product of the neurotic West, yet having essentials in common:

> To look upon it, is blindness
> To know it, is sickness.
> To worship it, is death
> To fear it is wisdom
> To resist it not, is redemption.   Sermo XI

> Throw in all the colours and the eyes might as well be blind;
> Throw in all the notes and the ears might as well be deaf;
> Throw in all the savours and the palate will taste nothing;
> To be forever chasing will lead to mania;
> To pursue always after something more valuable
> will spoil the character;
> So the wise man goes by need and not by appearance,
> And makes deliberate choice.   *Tao Te Ching* Chapter XII

The process of bringing the unconscious pole of a polarity to con-sciousness and suffering the resulting tension gives rise to a symbol to which Jung gave the name 'transcendent function', because of its quality of uniting the opposites. The ternary seems to have been for Jung very much a transitional step in the generation of a quaternary, as exempli-fied by the symbol as the resolution or transcendence of a paradox paving the way for the wholeness of the mandala.

> Among the various characteristics of the centre, one that has struck me from the beginning was phenomenon of the quaternity.

> ... we come, at least in my opinion, to the inescapable conclusion that there is some psychic element which expresses itself through the quaternity ... If

I have called the centre the 'self', I did so after mature consideration and a careful appraisal of the empirical and historical data.[10]

He found great significance, in *Answer to Job*, to the assumption of the Virgin Mary as adding a feminine fourth element to the patriarchal concept of the Trinity. Even in his typology one finds emphasis on the need to work on the third function as a means of bringing up the fourth – 'out of the three comes the four'.

In some ways this is strange. Certainly the introduction of a third element is often the disrupter of harmony – as the 'eternal triangle', sometimes resolved by bringing in a fourth. Yet for me it is pre-eminently the symbol of stability. A three-legged stool is stable, however badly made, whereas a four-legged chair or table is rarely stable: how often it is necessary to wedge one of the feet. The science of trigonometry is used for fixing positions on a map ('triangulation'), or the position of a ship on the ocean, or its distance from the shore. Musical harmony is based on the triad chord; three dimensions define an object in space. As I will come to suggest later, I believe the paradox may lie in the dimensions in which the symbol is viewed. Triangulation is a two-dimensional exercise. If one looks at the stool, on the other hand, there are really four sides involved in its stability – the floor makes the fourth. Viewing a man in two dimensions is rather like seeing him without his shadow, which gives him his solidity.

Jung is far from consistent in his expression of the term 'transcendent function'. Essentially any symbol arising from the unconscious has a transcendent function. In *Mysterium Conjunctionis* (CW 14), however, he uses this term seemingly in a special sense (cf. page 203) in relation to the uniting of pairs of opposites, which then express 'totality'. Pairs of opposites arranged in quaternary form are said to represent a conscious and differentiated totality. In this sense the symbol or transcendent function becomes a symbol of the Self. A speculation I find attractive (though it may strike horror in the minds of biologists) is the idea that a symbol may have a function in uniting the two halves of the brain.[11] Thus for Jung, the mandala, in which a quaternary structure, a differentiated totality imposed upon the All, is the supreme Self symbol.

> To the best of my experience we are dealing here with very important 'nuclear processes' in the objective psyche – images of the goal, as it were, which the

psychic process, being goal directed, apparently sets up of its own accord, without any external stimulus.[12]

More strictly, it seems that the presence of the second pair symbolises that for wholeness all polarities require resolution or transcendence. But Jung is quite open about the fact that it was the universal fascination of mankind with the quaternity which led him to his theories of its application to 'wholeness' and not the reverse (e.g. through a study of mandalas – see Note 2 above).

Viewed in two dimensions, there are two complementary, or opposed, (abstract) manifestations of a quaternary, each having numerous variations: the cross and the quadrilateral. Of these the upright cross and the upright square are the most usual, the diagonal cross and diamond (rotated square) being alternatives, each having (according to Cirlot) slightly different emphasis of meaning. The square has a material or earthly quality, and the cross a spiritual or heavenly quality. The cross is the subject of an extensive discussion in the Dream Seminars 1928-30.[13] Jung regards this symbol as arising as an abstract inner perception in the primitive brain:

> So Paleolithic man saw the absolutely abstract, a true ghost, and it made a tremendous impression on him.[14]

Jung attempts to justify this remark (relating to the 'sun wheel' found in Zimbabwe) by pointing out that geometric shapes do not appear in nature except perhaps in crystals. I think that he did not wish to detract from the undoubted wonder that primitive man should be capable of drawing such a design as a circle with two crosses dividing it (upright and diagonal). But to my mind this will not do. Primitive man was undoubtedly familiar with crystals (they were after all experts in stones, and quartz occurs commonly with flint), and would have seen refracted rays of light from them and from drops of water, and also the priceless sparks that generated their fire. Even more common would be instances of diffracted light – rays of light which appear when a strong light is viewed through a pinhole. This is a particularly common sight when the sun is seen through the leaves of a forest. I believe this is what St. Hubert saw between the antlers of the stag, in another example given by Jung of what was undoubtedly a Self symbol. To my mind it does not detract from the numinosity of a Self symbol that it has (or might have) an origin in nature, rather than something 'ghostly', but rather the contrary. Jung pointed out how precise was the observation of these people in their depiction of animals. It is strange that he quotes from Svendivogius without apparently seeing the analogy with diffraction:

> When a man is illuminated by the light of nature the mist vanishes from his eyes, and without difficulty he may behold the point of our magnet, which corresponds to both centres of the rays, that is those of the sun and the earth.[15]

There is also a correspondence with the eye, recognised in expressions such as 'a twinkle in his eye', 'star-crossed lovers', 'his eye shot daggers'. The iris of the eye is a clear circle, and the vivid expressions of the eye, especially in unconscious communication, certainly resemble a star or double cross (a strange expression) very like the sun-wheel referred to by Jung (above). The eye is included by Jung in his list of formal elements of mandala symbolism.[16] The origins of the cross will always be a matter of speculation and fantasy, and it is probably foolish to grasp at any single solution, since the origins of man were so widespread. What I have found striking is that the cross and star have the similarity that the lines or rays emanate from a central point, symbolic of origin, spiritual centre or soul. The eight-pointed star corresponds to both forms of the cross superimposed.[17] Jung quotes from Heraclitus: 'the soul is a spark of stellar essence', and from Meister Eckhardt: 'little soul spark'.

It is perhaps worth remarking that the wheel – e.g. the wheel of life in many Tibetan pictures – contains the centre, whereas the mandala need not, and often does not, though it is always at least implicit.

In Jung's words there is a 'premonition of a centre of personality'.[18] On the other hand, while the circle and cross are common in Christian symbology, the square is not. Rectangular windows in churches seemed first to have appeared in England with the 'new age' of Newton and Christopher Wren.

The relation of the centre to the periphery is dealt with in a very beautiful exposition of the individuation process to be found in *Psychology and Alchemy* (CW 12). Jung talks about a (sometimes spiral) circulation around the centre, 'like a shy animal, at once fascinated and frightened'. There is a comparison with other symbols, e.g. a spider in its web; also the idea of a magnet (the 'magnesia' of alchemy) 'capturing the processes of unconscious in a crystal lattice', of which more later.

It seems that the Self is the centre and also, in another sense, the whole, which includes the centre and all that is in the 'magic circle' which protects it like a womb or temenos – 'everything that belongs to the self – the paired opposites that make up the total personality'. It is compared to 'a conglomerate soul' in *Mysterium Conjunctionis*:

> Their [mandalas] basic motif is the premonition of a centre of personality, a kind of central point within the psyche, to which everything is related, by which everything is arranged, and which is itself a source of energy. The energy of the central point is manifested in the almost irresistible compulsion and urge to 'become what one is' … This centre is not felt or thought of as the ego but, if one may so express it, as the 'self'. Although the centre is represented by an innermost point, it is surrounded by a periphery containing everything that belongs to the self – the paired opposites that make up the total personality. This totality comprises consciousness first of all, then the personal unconscious, and finally an indefinitely large segment of the collective unconscious whose archetypes are common to all mankind … The self, although on the one hand simple, is on the other hand an extremely composite thing, a 'conglomerate soul' to use the Indian expression.[19]

(Actually, what Jung refers to as the 'conglomerate soul' may in Indian lore be a concept of 'group soul' in which different incarnations are combined – a rather different concept.)

So the circle is a dual symbol – the great All, and the 'small all' of the Self. The cross, or the star or 'double cross' also have a dual motif: they point to the centre, the nucleus, and they represent a summation of all the polarities of the 'personality' – presumably used to make sure the 'soma' is included with the 'psyche'. The square, by comparison to the 'wheel of life', seems to be present to anchor the spiritual within the material plane, to fix it, so to speak, once and for all within the crystal lattice and cease the endless *circumambulatio*, the spiral dance of the individuation process. The square is functionally tangential to the circle, although not always so depicted. The square implies also the two arms of the cross: it is quaternary, with opposite sides, and it implies the centre – the diagonals mark it even if they are not connected. It could be said to embody the cross.

But the cross without the square 'matrix' well represents the spirit unconnected to the earth. Crucifixion, to my mind, symbolises an unbalanced sacrifice of the material to the spiritual, quite different from the synergism of the mandala.

The symbol for the diamond (i.e. a rotated square, usually with corners stretched vertically) is not included by Jung in his list of mandala symbols[20] except perhaps implicitly as a quadratic temenos structure. This is a striking exception because the Tibetan mandala illustrated and discussed earlier focuses on the 'diamond thunderbolt' or *dorje*. The 'centre' list includes only the sun, star or cross. This exception is made good in his commentary on the Chinese alchemical text *The Secret of the Golden Flower*, to be discussed in the next chapter.

I find an association to the mandala in the (biological) cell, the building block of life, with its nucleus, semi-permeable envelope and other working components. How often is it drawn as a square with the nucleus at the centre, although its shape must change constantly? Plate 2d is a tracing taken from an electron microscope photo of a liver cell from a young chick, showing the nucleus at the centre of the space (more truly a volume) defined by the cell membrane.[21] Steven Rose comments:

> Yet it is these fixed patterns from the electron microscope that, as a result of the technology, form the basis for drawings in biology textbooks, and provide the conventional 'mind picture' of cells even for experienced biologists. So powerful is the technology that it becomes very hard to move beyond it, to think in three, let alone four dimensions.[22]

The next chapter will indeed move into three and even four dimensions.

NOTES

1 'The Kingdom of God', Francis Thompson.
2 The Shri Yantra is often described as a mandala, and illustrated as an example in Cirlot's Dictionary of Symbols. According to Dr Lauf (lectures at the Jung Institute) it is not a true mandala which should contain as essential elements both assymetry (normally vertical) and cosmological symbology (usually gates in the sides or corners). However there is considerable divergence of opinion as to the essentials of mandalas. There are a number of references in *Psychology and Alchemy* (CW 12) and *Mysterium Conjunctionis* (CW 14) to the quaternary structure of the mandala being equivalent to the *quaternio* of the alchemists (cf. page 15) but the structure referred to seems (as in Cirlot) to be a square, as in the comparison to the symbol for salt on page 245 of CW 14.
3 Cirlot, *Dictionary of Symbols*, Routledge.
4 *Ibid.,*
5 *Ibid.,*
6 Chungliang Al-Huang, *Quantum Soup*, Celestial Arts, 13.
7 *Ibid.,* 14
8 J.J. Clarke, *Jung and Eastern Thought,* Routledge, 45.
9 Robert Pirsig, *Zen and the Art of Motorcycle Maintenance,* Morrow 253f.
10 C.G. Jung, CW 12, 217-18.
11 M.A. Mattoon, *Jungian Psychology in Perspective,* Free Press, 138.
12 C.G. Jung, *ibid.*
13 C. G. Jung, *The Seminars,* Vol.1 *Dream Analysis,* 340f.
14 C.G. Jung, CW 9 Part 1, 48
15 *Ibid.*
16 C.G. Jung, CW 9 part 1 361.
17 *Ibid.,* 48.
18 C.G. Jung, CW 9 Part 1, 357.
19 *Ibid..*
20 *Ibid.,* 361.
21 From Steven Rose, *Lifelines*, Allen Lane, 61.
22 *Ibid.*

10

Three-Dimensional Symbolism: The Diamond

To see a World in a Grain of Sand,
And Heaven in a Wild Flower,
Hold Infinity in the palm of your hand,
And Eternity in an hour.[1]

1 Psychology

If thou wouldst complete the diamond body with no outflowing
Diligently heat the roots of consciousness and life.
Kindle light in the blessed country ever close at hand,
And there, hidden, let thy true self always dwell.[2]

S O BEGINS the text of the *Hui Ming Ching – The Book of Conscious-*
ness and Life, which, although ostensibly a secondary text in *The*
Secret of the Golden Flower, really explains the meditation, which
forms the principal text, in terms of the Tao. As both Wilhelm and Jung
recognised, the truth expressed in the Tao is psychological rather than
metaphysical. As mentioned in the previous chapter, I had come to this
conclusion on my own some months before completely re-reading *The*
Secret of the Golden Flower. But Jung (not surprisingly) had preceded me by
many decades. There is a whole world of difference here from the
medieval alchemists who recognisably projected unconscious truths into
mundane material:

> My admiration for the great Eastern philosophers is as genuine as my
> attitude towards their metaphysics is irreverent. I suspect them of being
> symbolical psychologists, to whom no greater wrong could be done than to
> take them literally.[3]

Richard Wilhelm adds in a footnote:

The Chinese philosophers – in contrast to the dogmatists of the West – are only grateful for such an attitude, because they also are masters of their gods.

Nowhere, so it seems to me, does Jung wax so eloquent about the individuation process as in his commentary on these texts. Well he might, for the Tao condenses all that is contained in his painfully unfolded experiential ideas on individuation in simple (if symbolic) and beautifully eloquent terms. When this text is compared with the garbled crudity of Western alchemy, fighting its bloody and painful way out of the mire of medieval dogma, it is as though a new dimension has been suddenly opened. And indeed this is itself symbolised by the completion of the Diamond Body, the Self symbol of the Tao, central to Chinese culture, *no longer in two dimensions, but three.*[4]

In his commentary Jung defends his Western point of view in rather whimsical terms, on the one hand anxious that his readers should not leave go of their roots in Western civilisation; on the other sad that the way of the East (which must in context mean the Tao, rather than Indian mythology) is apparently not open. He wrote this commentary in 1929, and things are now very different. Fritjof Capra has written *The Tao of Physics*, recognising that science is freeing itself from the chains not only of Newtonian physics, which was becoming true in Jung's lifetime, but also of Cartesian dogma, which forms the deterministic, time-dependent ground-plan of science as we were brought up to understand it. Perhaps the epitome of this lies in the second law of thermodynamics, that processes proceed in time from order to disorder.

> *The Moving Finger writes; and, having writ,*
> *Moves on: nor all thy Piety nor Wit*
> *Shall lure it back to cancel half a Line;*
> *Nor all thy Tears wash out a Word of it.*[5]

Lovelock and Margulis have pleaded that the planet should be approached scientifically as a living unit, rather than a multiplicity of unrelated systems. In *The Turning Point*, Capra writes:

In transcending the Cartesian division, modern physics has not only invalid-ated the classical ideal of an objective description of nature but has also challenged the myth of a value free science. The patterns scientists observe in nature are intimately connected with the patterns of their minds ...[6]

and, speaking of Jung's synchronicity concept:

Jung saw these synchronistic connections as specific examples of a more general 'acausal orderedness' in mind and matter. Today, thirty years later, this view seems to be supported by several developments in Physics ... physicists are now also making a distinction between causal (or 'local') and acausal (or non-local) connections.[7]

Fifty years would be closer to the mark than thirty. Compare Jung's perspective in 1930:

The type of thought built on the synchronistic principle, which reached its apex in the *I Ching*, is the purest expression of Chinese thinking in general. In the West, this thinking has been absent from the history of philosophy since the time of Heraclitus, and only reappears as a faint echo in Leibnitz. However, in the interim it was not extinguished, but continued to live in the twilight of astrological speculation, and remains today at this level.[8]

It may well be that history will show that Jung's greatest contribution to mankind has been his courageous exposition of the concept of synchronicity, for which he received so much support from Richard Wilhelm's translation of *The Secret of the Golden Flower*.[9] In saying this I do not mean to detract from his contribution to psychology. But psychology tends to be viewed 'from inside' and has inherited so much of the defensive conservatism of the magic circle of initiates with which the medical profession has been cursed. And his technique of analysis owed so much to his own personal genius, and indeed his religious sense and inspiration, the loss of which seems to be in danger of depriving his successors of optimism and vitality, and indeed their obligation as his heirs of continuing his Socratic goading of society. Jung certainly seems to have been under no illusions as to the importance of the principle of synchronicity:

Anyone who, like myself, has had the rare good fortune to experience in a spiritual exchange with Wilhelm the divinitory power of the *I Ching* cannot for long remain ignorant of the fact that we have touched here an Archimedean point from which our Western attitude of mind could be shaken to its foundations.[10]

This Socratic mantle is being carried, inevitably, in a one-sided unconscious warfare by societies such as Greenpeace. Happily there are also
more conscious scientists like Capra and Lovelock struggling to bring
some sense of order and balance into the political and scientific arenas.
Indeed there are now a number writers in different disciplines collab-
orating, many associated with The Scientific and Medical Network. But
the energy of most scientific thinking is still extraverted, projected, in
our Western neurosis, on to the world of 'things'. In a sense, this is 'star
energy', the masculine energy of the collective. What is lacking is what
Jung uncovered in Wilhelms's text *The Secret of the Golden Flower* – that
epitome of introversion, the Diamond Body. *Be still then and know that
I am God*, exhorts the psalmist (46.10).

The Tao can be seen as the pathway of individuation, the path of
unfolding awareness. Thus, according to Wilhelm:

> The Chinese character [for Tao] is made up of the character for 'head', and
> that for 'going' … 'Head' can be taken as consciousness (it is also the 'seat
> of heavenly light'), and 'to go' as travelling a way, thus the idea would be: to
> go consciously, or the conscious way.[11]

At the heart of the Tao is the concept of stillness; action through
non-action, *wu-wei*, projected by the Chinese alchemists into the exercise
of meditation. I find Jung's commentary particularly striking here:

> What did these people do in order to achieve the development that liberated
> them? As far as I could see they did nothing (*wu-wei* – action through non-
> action) but let things happen. As Master Lu-tsu teaches in our text, the light
> rotates according to its own law, if one does not give up one's ordinary occu-
> pation. The art of letting things happen, action through non-action, letting
> go of oneself as taught by Meister Eckhart became for me the key, opening
> the door to the way. We must be able to let things happen in the psyche. For
> us, this actually is an art of which few people know anything. Consciousness
> is for ever interfering, helping, negating, and never leaving the simple growth
> of the psychic processes in peace.[12]
>
> > The Tao does nothing,
> > and yet nothing is left undone.
> >
> > *Tao Te Ching* Chapter XXXVII, tr. Alan Watts

Wu-wei is at the heart of the Taoist ideal of behaviour, if one can so
express it. According to Alan Watts:

... the principle of 'non-action' (*wu-wei*) is not to be considered inertia, laziness, 'laissez faire', or mere passivity ... in the context of Taoist writings it [action] quite clearly means forcing, meddling, and artifice – in other words trying to go against the grain of '*li*' ... [*wu-wei*] is what we mean by going with the grain, rolling with the punch ... taking the tide at its flood and stooping to conquer.[13]

Jung shows, in the passage above, a side of his own process and perspective not often mentioned, and too often unrecognised by later exponents, who tend to emphasise a need to force conscious decisions inappropriate at least to the introvert needs of the second half of life. The Taoist conception of the 'void' can be understood as a striking parable of the relative values of extraversion and introversion and of the attitudes of the first and second half of life:

> Thirty spokes join the rim to the hub;
> By using the voids we have a carriage.
> Clay is formed into a pot;
> By using the void we have a container.
> Doors and windows are cut in walls;
> By using the voids we have a house.
> We have what exists;
> Yet it is the non-existent that makes it useful.
>
> *Tao Te Ching*, Chapter XI

Jung's comment in his recorded interview, 'Heart to Heart', to the effect that if no decision is made to correct a psychic need then fate will intervene, can be incompletely interpreted. One can procrastinate through fear or indolence, or one can make an act of faith, but either way it may be that the appropriate decision has not yet matured. To force a decision then might disturb one's fate inappropriately. This is a basic premise of the *I Ching*. The relation of faith and fate is beautifully expressed in Chapter VII of the *Tao Te Ching*:

> Heaven is inexhaustible and earth endures.
> They do not seek life for themselves,
> Yet in this way they live indefinitely.
> So the wise man abandons concern for his own safety,
> And behold it is ever before him;
> He gives no importance to his existence,

> And behold he is preserved.
> Surely it is because he is free from self-concern
> That he can accomplish what his heart desires.

A fictional Christian personification of *wu-wei* can perhaps be found in the character of Sybil in Charles Williams' fantasy *The Greater Trumps*. Indeed others among his characters seem to exhibit it, such as the archdeacon in *War in Heaven*. Among real people, Iris Murdoch comes to mind. I have already quoted Andrew Harvey's recollection in Chapter 7:

> She had no need to impress or prove anything, was an astonishing example of how to wear fame and assume the dignity of an elder, never for one second the *grande dame*. Her natural radiation stemmed from a powerful, peaceful, gentle wisdom, her journey an increasingly wide embrace from an increasingly private centre.

The goal of the Chinese alchemists' meditation was to bring into unity the concepts *hsing* and *ming*. These seem very difficult to translate due to the unconscious viewpoint of the Chinese and their pictorial calligraphy. Wilhelm and Jung use 'human nature' and 'life':

> The subtlest secret of the Tao is human nature and life (hsing-ming) There is no better way of cultivating human nature and life than to bring both back to unity.[14]

Hsing and *hui* are said by Jung to be *yang* principles, and *ming* to be *yin*. We know from the *I Ching* that the heavenly or divine is *yang* and earth is *yin*. When Wilhelm and Jung refer to 'life' in this context it must be rather as life (*ming*) within the earth, 'living nature'; 'human nature' (*hsing*) must be rather closer to the conscious or spiritual element in humanity than instinct, which the expression 'human nature' tends to suggest. In fact on page 103 Jung writes: 'The unity of these two, life and consciousness is the Tao,' although elsewhere consciousness is reserved for the Chinese *hui*:

> The unity of these two, life and consciousness, is the Tao, whose symbol would be the central white light ... This light dwells in the 'square inch', or in the 'face', that is, between the eyes. It is the image of the creative point, a point having intensity without extension, thought of as connected with the space of the 'square inch', the symbol for that which has extension. The two

together make the Tao. Human nature [*hsing*] and consciousness [*hui*] are expressed in light symbolism, and are therefore intensity, while life [*ming*] would coincide with extensity.[15]

The meditative image of the circulation of light mimics the movement of the sun around the earth. Here we have indeed the diamond and the star: the light of the star, the sun wheel, filling the square inch, the fixity, and completing the Diamond Body. This circulation is beautifully compared by Jung to the *circumambulatio*, the defining of the temenos or magic circle:

> The 'enclosure', or *circumambulatio*, is expressed in our text by the idea of a 'circulation'. The 'circulation' is not merely motion in a circle, but means, on the one hand, the marking off' of the sacred precinct, and, on the other, fixation and concentration. The sun wheel begins to run; that is to say, the sun is animated and begins to take its course, or, in other words, the Tao begins to work and to take over the leadership. Action is reversed into non-action; all that is peripheral is subjected to the command of what is central. Therefore it is said: 'Movement is only another name for mastery.' Psychologically, this circulation would be the 'turning in a circle around oneself', whereby, obviously, all sides of the personality become involved. 'They cause the poles of light and darkness to rotate,' that is, day and night alternate.
>
> > *Es wechselt Paradieseshelle*
> > *Mit tiefer, schauervoller Nacht.*
> > *Faust* ('The radiance of Paradise alternates with deep, dreadful night.')
>
> Thus the circular movement also has the moral significance of activating all the light and the dark forces of human nature, and with them, all the psychological opposites of whatever kind they may be. It is self-knowledge by means of self-incubation (Sanskrit *tayas).*

One is reminded of the Buddhist monks tramping their *stupas*, or the old English practice of 'tramping the bounds' of the estate to create the 'home'. It is the definition of the 'small All', the microcosm of the personality. Compare the original text:

> The circulation is fixation. The light is contemplation. Fixation without contemplation is circulation without light. Contemplation without fixation is light without circulation. Take note of that![16]

Clarke writes about this:

The text suggests the idea of transformation, not as a linear, but as a circular process, in effect like the hermeneutical circle we noted earlier, in which a rotation from one pole to another, from light to darkness, brings about a progressive sense of integration and completeness. This circular movement 'has the moral significance of activating the light and dark forces of human nature' (CW 13.39), a procedure in which all sides of the personality come into play, and which leads to self-knowledge. The 'circumambulation of the self' was a favourite image of Jung's that he used to express the dynamic process of self-discovery whereby the ego is seen to be contained in the wider dimensions of the self, the self being both the centre and the circumference of the psychic life.[17]

What a comparison with modern life! The original unity, at birth, it is said, 'dwells in the square inch (between the eyes)'. The 'conscious spirit' is said to dwell below in the heart. These are brought into unity, when the circulation proceeds of its own accord, under the control of the centre. And so we get to the description of the ultimate vision:

> Without beginning, without end,
> Without past, without future.
> A halo of light surrounds the world of the law.
> We forget one another, quiet and pure,
> altogether powerful and empty.
> The emptiness is irradiated by the light of the heart
> and of heaven.
> The water of the sea is smooth and mirrors the moon
> in its surface.
> The clouds disappear in blue space;
> the mountains shine clear.
> Consciousness reverts to contemplation;
> the moon disc rests alone.

This is described by Jung as 'detachment of consciousness from the object', which sounds like pure introversion, and yet is a picture of the individuated condition. Wilhelm remarks that the Chinese character *ho*, translated as 'individuation', is written with the symbol of 'energy' inside an 'enclosure'. The text states:

> After a man has the one sound of individuation behind him, he will be born outward according to the circumstances and until his old age he will never look backward.[18]

Jung says of the vision:

> By understanding the unconscious we free ourselves from its domination...
> This description of fulfillment pictures a psychic state which perhaps can be
> best characterised as a detachment of consciousness from the world and a
> withdrawal of it to an extra-mundane point, so to speak. Thus consciousness
> is at the same time empty and not empty. It is no longer preoccupied with
> the images of things, but merely contains them ... The magical claim of
> things has ceased because the original interweaving of consciousness with
> the world has come to an end. The unconscious is no longer projected, and
> so the primal 'participation mystique' with things is abolished.[19]

Jung's writings on 'participation mystique' are not easy to assimilate.
The attention given by him to this important concept in the foreword to
The Secret of the Golden Flower requires for its clearer understanding some
exploration and I have felt it best to give a discussion of this topic in an
appendix (page 163). Essentially it is a form of psychic identity with an
object of veneration. In a primitive, who can be defined for present
purposes as someone lacking the clear conscious/unconscious division
present in a Western adult, and who lives his life in a child-like imaginary
world, this identity is normal and in general beneficial. The great
examples of this are perhaps those described by Jung's disciple Sir
Laurens van der Post in his studies of the Bushmen of the Kalahari:

> ... there was a constant traffic of meaning between the stars and starlight,
> and themselves and their spirit; they participated deeply in one another's being
> ... They called Jupiter 'the Dawn's Heart'. He was so close to them, so much
> a person, that he once came down to earth and took a wife ... the Lynx ...[20]

However, in a conscious Western adult where 'imaginary' contents are
suppressed, the resulting projections can take the form of obsessions and
need to be recognised. Jung bemoans the possession of the 'unconscious'
Western psyche by the archetypal remnants of the old gods and demons
which he calls 'de-deification'[21] and extols the virtue of consciousness by
which Western man individuates by giving recognition to the gods and
demons ('re-deification'). This re-deification he found in the meditations
of the Chinese alchemists, the amalgam of the diamond and star.

At the front of *The Secret of the Golden Flower* is a picture of a Lamaist
vajri-mandala (Plate 2a) which Jung discusses in CW 9, Part I:

The mandala shown here depicts the state of one who has emerged from contemplation into the absolute state. That is why representation of hell and the horrors of the burial ground are missing. The diamond thunderbolt, the *dorje* in the centre, symbolizes the perfect state where masculine and feminine are united. The world of illusions has finally vanished. All energy has gathered together in the initial state.

The symbol of the *dorje* has impressed me profoundly and may have contributed to the original impetus for my ruminations on the diamond and star. For here they were both united: the energy of the cosmos, the world of the stars, and the, fixity and endurance of the most perfect substance of the earth. Jung continues:

> The four *dorjes* in the gates of the inner courtyard are meant to indicate that life's energy is streaming inwards; it has detached itself from objects and now returns to the centre. When the perfect union of all energies in the four aspects of wholeness is attained, there arises a static state subject to no more change. In Chinese alchemy this state is called the 'Diamond Body', corresponding to the 'corpus incorruptibile' of medieval alchemy, which is identical with the 'corpus glorificationis' of Christian tradition, the incorruptible body of resurrection ... This mandala shows, then, the union of all opposites, and is embedded between *yang* and *yin*, heaven and earth; the state of everlasting balance and immutable duration.
>
> For our more modest psychological purposes we must abandon the colourful metaphysical language of the East. What yoga aims at in this exercise is undoubtedly a psychic change in the adept. The ego is the expression of individual existence. The yogin exchanges his ego for Shiva or the Buddha; in this way he induces a shifting of the psychological centre of personality from the personal ego to the impersonal non-ego, which is now experienced as the real 'Ground' of the personality.[22]

This passage strikes a rather discordant note within Jung's concept of 'wholeness' contrasted with 'perfection'. Although lip service is paid to wholeness, the passage *in toto* suggests rather a state of perfection. Perhaps the saving grace lies in the word 'balance', which would reconcile with the Tao. The diamond, like any crystal, is essentially energy in a particularly harmoniously balanced three-dimensional lattice, in contrast to the square and circle, which are essentially abstract and 'unnatural'. If one had ears tuned to its vibrational frequency, one would hear a pure note, the *ho*. This would symbolise organic life, not death, which is the opposite pole of the pre-eminently static state of perfection – the *corpus glorificationis* as traditionally understood. In all these references to the individuated state one must pay due respect to Jung's insight into the psychology of alchemy, mudded as the traditional understandings are with the confused conceptions of the middle ages.

The passage of course represents a mystical experience which is more fully discussed in Appendix I. While in a metaphysical sense this represents an ideal situation, a total loss of illusion for the Hindu or nirvana for the Buddhist, in psychological terms consciousness is still present. If it were not so, if consciousness were overwhelmed, this would imply a state of identity with the divine, gross inflation, and obsessive infallibility. This type of mystical experience may be expressed as a confrontation with the light. Too great a light and one is blinded, as in the cry of the soul in *The Dream of Gerontius:* 'Take me away!'

The overwhelming power of the unconscious can, at the other pole, be experienced in a confrontation with the dark, the terrifying images of the nightmare. Whether light or dark, there is here a narrow borderline which Jung himself experienced and which is best, perhaps, described in *Memories, Dreams, Reflections:*

> In order to grasp the fantasies which were stirring in me 'underground,' I knew that I had to let myself plummet down into them, as it were. I felt not only violent resistance to this, but a distinct fear. For I was afraid of losing command of myself and becoming a prey to the fantasies – and as a psychiatrist I realised only too well what that meant. After prolonged hesitation, however, I saw that there was no other way out. I had to take the chance, had to try to gain power over them; for I realised that if I did not do so, I ran the risk of their gaining power over me. A cogent motive for my making the attempt was the conviction that I could not expect of my patients something I did not dare to do myself. The excuse that a helper stood at their side would not pass muster, for I was well aware that the so-called helper – that is myself – could not help them unless he knew their fantasy material from his own direct experience ...[23]

Jung let himself plunge into the depths and had some vivid experiences culminating in the image of an outpouring of blood, which at the time he associated with the Great War which was taking place (or imminent). However, in a dream which followed some days later, he felt himself impelled to kill an image of Siegfried, which made a powerful impression:

> After the deed I felt an overpowering compassion, as though I myself had been shot: a sign of my secret identity with Siegfried, as well as the grief a man feels when he is forced to sacrifice his ideal and his conscious attitudes. This identity and my heroic idealism had to be abandoned, for there are higher things than the ego's will and to these one must bow.[24]

The dangerous near-identity with the hero image was dissolved. What more can we associate with the symbol of the diamond? We know that the diamond is structurally crystalline, balanced energy in three-dimensional form, four if we include the dimension of time, since it is constantly vibrating. We know that it is close to being 'incorruptible': it is clear, and it is hard. Jung has pointed out that it is formed of carbon, the principle element of organic chemistry, the chemistry of living matter, and that has four valencies and occurs also in a black form. A good start, one might say. But what else do we know?

NOTES

1 'Auguries on Innocence', William Blake.
2 Richard Wilhelm, *The Secret of the Golden Flower*, 69.
3 *Ibid.*, 129.
4 The symbol is more employed by the Taoist alchemists, later than Lao Tsu.
5 *Rubaiyat of Omar Khayyam*, Edward Fitzgerald.
6 Fritjof Capra, *The Turning Point*, Flamingo, 77.
7 *Ibid.*, 400.
8 Richard Wilhelm, *ibid.*, 143.
9 *Ibid.*, xiii (Foreword to the second German edition).
10 *Ibid.*, 140.
11 *Ibid.*, 97.
12 *Ibid.*, 93.
13 Alan Watts, *Tao – The Watercourse Way*, Pelican, 75, 76.
14 Richard Wilhelm, *ibid.*, 69.

15 *Ibid.*
16 *Ibid.,* 25.
17 J.J. Clarke, *Jung and Eastern Thought*, Routledge, 85.
18 Wilhelm, *ibid.,* 32.
19 *Ibid.,* 122.
20 Sir Laurens van der Post, *The Creative Pattern in Primitive Africa*, Eranos Lectures No 5, 17, 18.
21 C.G. Jung, CW 18, 1364f.
22 Jung, CW 9 Part I, 358.
23 C.G. Jung, *Memories, Dreams, Reflections*, Collins, RKF, 172.
24 *Ibid.,* 174.

11
Three-Dimensional Symbolism:
The Diamond

2 Chemistry

Stately Spanish galleon coming from the Isthmus,
Dipping through the tropics by the palm-green shores,
With a cargo of diamonds,
Emeralds, amethysts,
Topazes and cinnamon and gold moidores.

Dirty British coaster with a salt-caked smoke stack,
Butting through the Channel in the mad March days,
With a cargo of Tyne coal,
Road-rail, pig-lead,
Firewood, iron-ware, and cheap tin trays.[1]

Carbon

I HOPE THAT this chapter will re-awaken our jaded sense of wonder to the miracle of carbon. Much of what follows will be well known to many with a background in science but I hope will nevertheless be stimulating to them as to many others whose education has led them along other directions.

Carbon is the magic element on which life itself depends. It is unique in so many respects among the elements. Firstly, it is 'tetravalent', meaning it can form compounds with four atoms of elements with a valency of one, like hydrogen, or two atoms of those with a valency of two, like oxygen, and so on. It is supremely non-metallic and forms compounds by a process of 'sharing' valences (co-valency). It has four unfilled or 'receptive' sites in its outer shell of electrons, and similar filled or 'donor' sites. When combined, its own four electrons are bequeathed in return for four more electrons from its satellite elements to form a stable eight-electron shell which is a 'commonwealth'. It cannot relinquish or capture electrons to become a charged ion; nor can it exist in a monatomic state. This is related to its position midway between the satisfied electronic states of two inert elements, who, with eight electrons in their outer shells, are chemically 'dead', like argon. Therefore symbolically it is supremely 'alive'. It is supremely 'relational'. It mimics vividly the unsatisfied condition of man who is unrelated, who longs for a wholeness which can only be satisfied by a process of sharing, whether with a woman, or society, or through individuation. There are no (sane) lone wolves in the human world for as St. Paul says 'it is better to marry than to burn'.

> Man is weak, and therefore is communion indispensable. If your communion be not under the sign of the Mother, then it is under the sign of the Phallos. No communion is suffering and sickness. Communion in everything is dismemberment and dissolution.
>
>   C.G. Jung, Sermo V

The Element
Carbon exists in two elemental forms, diamond and graphite, in both of which the element forms a compound with itself, in which there is no residual tension between oppositely charged particles. The electrons in the outer rings are shared 'co-valently'.

A most remarkable fact about carbon is that the four valencies are (when most stable) equi-directionally arranged in three dimensions. The valences extend to the points of the solid known as the tetrahedron (four-pointed) which is quite difficult to draw in two dimensions (Plate 12).

Each of the four sides is an equilateral triangle. The solid is rarely seen and many people are unfamiliar with it. However it does occur in nature, e.g. in certain seeds like the beech nut. It is the most elementary or simplest possible solid, just as the triangle is the most elementary or simplest possible rectilinear closed figure.

The Structure of Diamond
As one might perhaps predict from its triangular roots, it also provides the closest known and therefore the most stable arrangement of atoms in a crystal lattice. In the diamond, all the atoms are identical – carbon

– and are tetrahedrally arranged into a tight structure which explains its extraordinary properties (Plate 13).

Strangely, this structure also exhibits cubic symmetry.

Diamond is phenomenally stable, hard and clear. Unlike its allotrope, graphite, it does not conduct electricity although some synthetic diamonds have been made into semi-conductors by 'doping' with other elements. It burns only at very high temperatures. Strangely its composition was discovered at the time of the French Revolution by the aristocrat Lavoisier, who managed to burn it with a magnifying glass and identified the resulting carbon dioxide. It seems that all the 'maxima' imaginable might be attributed to it. It has the highest refractive index known, which accounts for its spectacular brilliance. As a chemical entity it is unique.

Symbolically, what can one say about this unique structure? Words begin to fail. It combines the fixity of the three with the completeness of the four, the magical seven. It is not merely an abstract construct like the square but a physical entity. As a symbol of the Self it seems without parallel. It adds a whole dimension to alchemical symbolism.

The Structure of Graphite

To reject that part of the Buddha that attends to the analysis of motor cycles is to miss the Buddha entirely.[2]

Diamond is the light, spiritual, conscious pole of the element carbon, whose chthonic form is graphite. Carbon might therefore stand symbolically for the ego, the *prima materia* of alchemy, for man/woman, holding the tension of the opposites; perhaps at a symbolic mid-life crisis where the conscious/unconscious split is at a maximum; or simply at any stage in the life cycle, somewhere between the conscious and unconscious.

Graphite is generally considered to be the only other crystalline condition of carbon. So-called 'amorphous' carbon, soot, which can be obtained as a fine powder from reacting gases, is generally considered to be microcrystalline graphite. In graphite the energy pattern of the outer electron shell is distorted into a flat ring. There is the same 'commonwealth' of electrons in each ring, but there are fewer than the number needed to satisfy the requirements of tetrahedral bonding. Three of the bonds are 'double bonds'. This can be tolerated because of the peculiar power of the ring structure to 'distribute the inequality'. The rings appear hexagonal, so that graphite forms monatomic flat layers like sections through a honeycomb (Plate 13).

There is almost no binding between the layers which are attracted together by weak 'van de Wahl's forces' and readily separate and slide over one another, giving graphite its slippery, lubricating feel. It is often used as a substitute for oil.

Many carbon compounds, in general those derived from coal, like benzene, share this 'unsatisfied' ring. So Kekule's prophetic dream described in the introduction, which led to the discovery of the hexagonal ring of benzene, might be said to refer more profoundly to the structure of carbon itself.

Graphite is very opaque to light – 'black as soot'; it is porous; it burns; it conducts electricity; it fragments; it lubricates. Unlike the diamond, which is useless except for special purposes, it is a very practical, useful substance. It provides energy (coal); it is the most widely used dark pigment; although it discolours it also draws, as its name implies.[3] Although graphitic carbon is frangible when exposed, it can be exceedingly strong when embedded. Carbon fibres can be made by removing the elements of water from cellulose, which consists of 'chains' of connected carbon rings. When embedded in a 'matrix' of resin they act as a reinforcement which is stronger, weight for weight, than glass or steel. It can be 'activated' by exposure to heat in a vacuum to act as a catalyst or absorptive medium. Its absorptive powers are cleansing: it will absorb noxious fumes in a kitchen or poisons in the intestines. This will eventually poison the carbon, but it can be re-activated.

Like diamond, it is very stable and reacts only with difficulty. Somewhat strangely, graphite is more stable thermodynamically than diamond, which will gradually revert to this form, albeit inappreciably slowly at normal temperature, as can be seen in Plate 14.

The phase diagram shows that, thermodynamically speaking, diamond is only stable in solid form at very high pressures – around twenty atmospheres. It relies for its stability on the very high activation energy needed for conversion from graphite, as shown in the energy diagram. Sufficient energy must be imparted to 'lift it over the hump'. Industrial diamonds can be made using catalysts which reduce the energy needed, but at the cost of discolouration. Diamonds are formed in 'pipes' of molten volcanic material moved to the surface under pressure and rapidly frozen. If the energy diagram is imagined in three dimensions diamond can be seen to occupy a verdant plateau surrounded by almost inaccessible mountain ranges – a veritable Shangri La.

Carbon Compounds
The number of carbon compounds known is almost too high to be counted and chemists are discovering or synthesising more every day. The whole of life on the planet revolves around the reactions of carbon compounds – but, strangely enough, these reactions carry on among the compounds themselves without reference back to the unreactive elemental forms. It is as though diamond and graphite stand aside like fond parents viewing the mysterious antics of their children.

The structure of carbon compounds varies according to whether the 'satisfied' tetrahedral configuration visible in the diamond is used or whether, e.g., flat rings of the graphite type are present. The simplest carbon compound is methane in which each tetrahedral bond is attached to a hydrogen atom (Plate 15).

But as shown in its elemental forms, carbon atoms can attach to each other in chains, rings and many combinations of the two. Not only 'double bonds' may be present but also 'triple bonds', as in acetylene. Many compounds contain rings of carbon atoms; some, like benzene which gave such trouble to Kekulé, have the flat, graphitic structure. In benzene the spare bonds of the ring, instead of being shared with those of neighbouring graphite rings, are shared with those of individual hydrogen atoms. These compounds are classified as 'aromatic', due to the distinctive smell present in many of them, mostly derived from coal tar, like the first dyestuffs. The first group are classified as 'aliphatic'. Aliphatic rings are present in many compounds associated with life. For example cellulose which forms the main constituent of plants, has a hexagonal ring which is not flat but bent into a U or S into which the tetrahedral bonds naturally fall (Plate 15).

In the diagram all the extending arms represent bonds connected to hydrogen atoms. The great class of 'carbohydrates' in which the elemental constituents of water are present are made up of ring shaped glucose molecules connected together. Plate 15 shows the formula for sucrose as normally written, made up of a molecule of glucose and one of fructose connected together. But the rings are not really flat.

Plate 16 shows how the chains in starch and glycogen, based mainly on the U or 'boat' structure cannot support long chains and are broken up; it can be seen that even the S or 'chair' structure does not make for straight line polymers. The picture of cellulose below shows the kind of 'tangled web' which gives cotton its softness and absorbency compared to nylon.

Of course, it is important to life that carbon can attach itself to other, mainly non-metallic, elements. In the compounds above divalent oxygen as well as monovalent hydrogen is present. When we come to proteins, we find that divalent sulphur can substitute for oxygen and trivalent nitrogen is an important constituent; pentavalent phosphorus is an ingredient of DNA and important in energy transfer. Metals are also present in carbon compounds and essential to life, such as iron in blood, magnesium in chlorophyll, calcium in bones. The versatility of carbon is really extraordinary. Perhaps the most unusual compounds are the hollow spheres which have been synthesised from the element itself. It was found that insertion of a layer of a graphite variation containing pentagonal rings between two layers of graphite forced a spherical distortion. After the designer of the Geodesic Dome they have been given the name 'Buckminster Fullerite.' One such compound has sixty carbon atoms (Plate 17).

Although diamond is the closest packed naturally occurring carbon structure, synthetic chemists have imagined a structure containing carbon and nitrogen atoms, C_3N_4, as being even more densely packed, since nitrogen is a smaller atom and would take up less space: This might in the future prove a competitor for diamond at least in industry.

Finally, all of living matter depends to one extent or another on the ability of carbon to react. Although its elementary forms themselves show a strange reluctance to react, rather like Abraham, its family of dependent compounds are as numerous as the sand on the seashore. It well symbolises unconscious man, and throws some new imagery on consciousness.

The Cycle of Terrestrial Life

I believe it to be a fact, a remarkable fact if true, that the carbon within the earth all seems to have originated in living matter. Certainly coal is the compressed debris of plants, passing through the stages of compost and peat. Mined graphite seems to be the product of further degeneration of coal. Finally diamond seems to be graphite which has been tortured by volcanic action, in which it has been heated to enormous temperature in the absence of reactive elements and compressed by infolding of the matrix rocks to enormous pressures. Science has not yet been able to emulate this process except to the extent of producing microscopic black crystals of adamantine material passing under the uneuphamous title of 'industrial diamonds'.

The origin of carbon, and therefore of life on earth, is strangely 'extra-terrestrial'. It would not be an unreasonable flight of the imagination to ascribe it to stellar debris, the breath of falling stars.

The source of carbon for plant life, and so indirectly for animal and human life is not the earth, but rather the carbon dioxide of the air, through the magical daily cycle of photosynthesis in which the necessary energy comes from the light of the sun, our most important star, giving us at the same time the oxygen we breathe. The supply of carbon dioxide is constantly being renewed by combustion of organic matter, and unfortunately accelerated beyond our tolerance level by the burning of fossil fuels and of huge tracts of the equatorial rain forest, which of course has a 'double bind' of increasing the supply and reducing the consumption. Eventually we will bake in a solar oven, or perhaps asphyxiate if we are not first drowned in the rising sea as the ice caps melt.

This fate is not something to be dismissed for the distant future. Unless human nature experiences a miraculous change we may be one of the last surviving generations.

So the cycle is complete: from the stars to the air; from the air to the plants; from the plants to the animals, to man, to the earth; and from the earth, and within man to the diamond, capturing and fracturing the starlight into the numberless gods of the *VII Sermones ad Mortuos*, and the numberless possibilities of the Tao.

> The Tao is like a well:
> Used but never used up.

It is like the eternal void
Filled with infinite possibilities…

The Tao is called the Great Mother
Empty yet inexhaustible,
It gives birth to infinite worlds.

It is always present within you.
You can use it any way you want.[4]

NOTES

1 'Cargoes', John Masefield.
2 Robert M. Pirsig, *Zen and the Art of Motorcycle Maintenance*, Morrow, 83.
3 From grafein to write.
4 Lao Tsu, *Tao Te Ching*. Quoted by Alan Watts in *What is Tao*, in selected chapters.

12

Twilight

Sunset and evening star,
And one clear call for me …

Twilight and evening bell,
And after that the dark[1]

ENNYSON'S POEM evokes the mystery of the threshold of death, a time of recollection and resignation symbolised and personified by the evening star. But the evening star is also the morning star, the symbol of hope:

Brightest and best of the sons of the morning,
Dawn on our darkness and lend us thine aid,
Star of the East the horizon adorning
Guide where our infant Redeemer is laid.

The Epiphany hymn extols the story of the three wise men: 'The star which they saw in the East went before them, till it came and stood over where the young Child was.' Christian imagery seems to have picked up on the morning star, next portrayed as a symbol of the Messiah himself:

Arise O Morning Star, arise and never set!

This is not surprising. It takes an effort, now, to loose the banalities surrounding Christian worship and imagine what the emergence on the scene of the person of Jesus must have been like in a society cowering under the cruelty of Rome, the cynicism of Herod and the anal hypocrisy of the religious authorities. The evening star could have been symbolised (but was not to my knowledge) by John the Baptist. Midnight is portrayed by the feast of Advent:

'Wachet auf,' ruft uns die Stimme
Der Wächter sehr hoch auf der Zinne,
Wach auf du Stadt Jerusalem!
Mitternacht heißt diese Stunde!'[2]

So far the Star has been viewed predominantly as the sun, the bright Apollonian symbol of consciousness. But the liminal realm of twilight is its birthplace. As T.S. Eliot puts it:

> Between the idea
> And the reality ...
> Between the potency
> And the existence ...
> Falls the Shadow.[3]

The Shadow is left to our imagination. It can be seen as despair, or the encounter with the repressed energies of shame and guilt, in a Jungian sense, or more simply as the twilight obscurity where creative symbols emerge. In all there is truth.

Although Venus is more commonly regarded as the evening and morning star, Mercury shares the attribute, but is more difficult to see, always close to the horizon, and to the sun from which it deviates never more than 27 degrees (Plate 18a). Venus is the brightest in the night sky and often well above the horizon (Plate 18b). Both are mythical psychopomps, messengers between the conscious and unconscious, between man and woman, mankind and nature, mankind and God. Mercury, badge of the Corps of Signals, operates more on the verbal plane; Venus, personification of feminine beauty, on the sexual. But Venus, associated by Jung with the *anima* and by the alchemists as the mother of the alchemical Mercurius[4] can be regarded as generating psychic energy and Mercury as channelling it.[5] Mercury is the Greek God Hermes, the trickster who forces the intrusion of the unconscious with his games, those 'Freudian slips'; Venus is, of course, the Greek Aphrodite, Goddess of love.

These stars reign throughout the night, the time of reintegration, of dreams, where the unconscious prevails, gradually taking over, and gradually loosing its hold as the rising sun begins to dominate. Although stars in the mythical sense they are, of course, planets, reflecting the light of the sun so that in the twilight unconscious ideas are not driven out but can emerge in an extension of consciousness.

What takes place between light and darkness, what unites the opposites, has a share in both sides and can be judged just as well from the left as from the right, without our becoming any the wiser; indeed we can only open up the opposition again. Here only the symbol helps, for, in accordance with its paradoxical nature, it represents the *'tertium'* that in logic does not exist, but which in reality is the living truth … The *proteum mythologem* and the shimmering symbol express the processes of the psyche more trenchantly and, in the end, far more clearly than the clearest concept; for the symbol not only conveys a visualization of the process but – and this is perhaps just as important – it brings a re-experiencing of it, of that twilight which we can learn to understand only through inoffensive empathy, but which too much clarity only dispels.[6]

The liminal can be seen as a gateway, symbolic of initiation, an adolescence where the encounter of childhood ideals with the base realities of an imperfect world (the Shadow) must take place before an adult consciousness can emerge. A dream image of this gateway is shown in Plate 19. There seem to be two thresholds: in the first, at sunset, the obsessions of the old state lose their power and the psyche opens to receive whatever may arise from the depths. In the intervening 'night of the soul' the psyche is faced with the shadow and the archetypes of the Self. This can be seen as an encounter with the enlightening properties of Lucifer, vividly portrayed by Baudelaire in his *Fleurs du Mal*:

> Tête-à-tête sombre et limpide
> Qu'un coeur devenu son miroir!
> Puits de Vérité, clair et noir
> Où tremble une étoile livide,
>
> Un phare ironique, infernal
> Flambeau des grâces sataniques,
> Soulagement et gloire uniques,
> – La conscience dans le Mal![7]

Baudelaire returns to the theme in his 'Hymn to Beauty', clearly a reference to Venus:

> Tu contiens dans ton oeil le couchant et l'aurore;
> Tu répands des parfums comme un soir orageux;
> Tes baisers sont un philtre et ta bouche une amphore
> Qui font le héros lâche et l'enfant courageux.

De Satan ou de Dieu, qu'importe? Ange ou Sirène,
Qu'importe, si tu rends, – fée aux yeux de velours,
Rythme, parfum, lueur, ô mon unique reine! –
L'univers moins hideux et les instants moins lourds?[8]

Then, at the dawn, the experiences of the night are resolved into symbols resonant of a new consciousness.

The process of initiation is given the power of ritual in primitive societies. The anthropologist Robert Pelton distinguishes the three phases:

> The main aspect of a primitive rite of passage is characterised by three phases: 1. Separation, 2. Margin or limen and 3. Aggregation. The first phase comprises symbolic behaviour involving detachment of the individual or group, either from an earlier fixed point in the social structure or a set of cultural conditions (a state). During the intervening liminal period, the state of the ritual subject (the passenger) is ambiguous, he passes through a realm that has few or none of the attributes of the past or the coming state; in the third state the passage is consummated.[9]

In any initiation there is an element of sacrifice. There must be, as a prerequisite, a willingness to let go of earlier conceptions, and this is experienced as pain or grief and may be accompanied by shame. It is a small death in which the suffering, which is accepted and not repressed may – or may not – hold an expectation.

The twilight zone, represented by Pelton as the limen, faces the initiand not only with human symbols but those of Gods and nature:

> Liminality is replete with symbols quite explicitly relating to biological processes human and non-human, and to other aspects of the natural order. In a sense, when man ceases to be master and becomes the equal or fellow of man, he also ceases to be master and becomes the equal or fellow of non-human beings.[10]

One might add with Richard Wilhelm that he ceases to be subject to the Gods and becomes their master.[11] In effect, in joining his community, he becomes related, as does his community, to all beings.

Chungliang Al-Huang has interesting things to say about the gate in *Quantum Soup*.[12] Plate 20b reproduces his drawing of the Chinese character *men*, meaning gate. He projects the ideas of the gate showing both a *way in* and a *way out*. He also points to the first chapter of the

Tao Te Ching which speaks of the gateway to the mysterious void, the source of all possibilities, and its association with Genesis 'and the earth was without form and void, and darkness was upon the face of the deep.' Certainly the symbol of gate or gateway proliferates with meanings. But for me, perhaps as a result of my dream, it pre-eminently suggests a passage *through* to some new situation.

I was delighted to see Al-Huang inviting the reader to play with the hexagrams:

> Group them [the lines and strokes] in twos and threes and in fours and sixes. See the relationships. Look to them for your message. Allow your own wisdom to be triggered by these symbols.[13]

I accept! There are several passages in the *I Ching* which appear to relate to an initiation process and include hexagrams resembling gateways as depicted in Plates 19, 20a or 20b.

To start with, the trigram *Ken* (mountain or stillness) with an unbroken line covering two broken lines suggests it and could be taken as the psychic centring at the heart of any initiation. But as Chungliang Al-Huang points out elsewhere, the mountain is also 'home':

> Mountain (*Shan*) [an alternative character] is your home base, not the next 'Everest' you must climb! Your mountain is your summit and your bedrock To return and return again, home.[14]

The hexagram Standstill (Heaven over Earth) which is opposite to that of Peace can well represent a twilight situation. The preceding hexagram is indeed Peace, a position of static harmony, while the succeeding hexagram is Fellowship with Man. Standstill or Stagnation is, according to Wilhelm, a period of withdrawal, where the dark forces are within and can well be compared to a twilight period.

Peace   *Standstill*   *Fellowship*

The sequence Peace-Standstill-Fellowship appears at positions 11, 12 and 13 of the 64-hexagram sequence. At 22 to 25 there is a comparable sequence of Grace-Splitting Apart-Turning Point-Innocence or Unexpected. Grace, similarly to Peace, is a static period of harmony but with more transitory or fragile associations. The hexagram for Splitting Apart, with the Earth below the Mountain again resembles a gateway.

At Turning Point the hard line returns to a position at the bottom of the hexagram, so that the Earth is now covering Thunder or The Arousing. The hexagram might be a step or threshold.

Finally in Innocence, Heaven now replaces Earth above Thunder or the Arousing, indicating nature directed by or in accord with the spirit.:

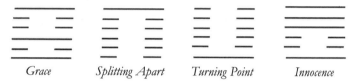

Grace   *Splitting Apart*   *Turning Point*   *Innocence*

Strangely enough, this sequence immediately follows the earlier sequence: Work on what has been Spoiled-Approach-Contemplation-Biting Through, which leads to Grace. In this more extended sequence, a threshold (Approach) seems to precede the gateway (Contemplation).

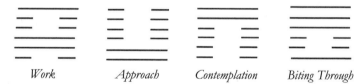

Work   *Approach*   *Contemplation*   *Biting Through*

So Contemplation is sandwiched between two intermediate states. In this case (Contemplation), Wilhelm illustrates what I have termed a gateway but calls it a tower:

> A slight variation of tonal stress gives the Chinese name for this hexagram a double meaning. It means both contemplating and being seen, in the sense of being an example. These ideas are suggested by the fact that the hexagram can be understood as picturing a type of tower characteristic of ancient China. A tower of this kind commanded a wide view of the country; at the same time, when situated on a mountain, it became a landmark that could be seen for miles around. Thus the hexagram shows a ruler who contemplates the law of heaven above him and the ways of the people below, and who, by means of good government, sets a lofty example to the masses.

For once I must disagree with Wilhelm. I have not visited mainland China, but throughout Japan such gates are situated at the entrances to temples, marking for the pilgrim the transition from the profane to the sacred temenos. An example is shown in Plate 20a.

Following 'The Judgement' Wilhelm says:

> The sacrificial ritual in China began with an ablution [*Approach*] and a libation by which the Deity was invoked, after which the sacrifice [*Biting Through*] was offered. The moment of time between these two ceremonies [*Contemplation*] is the most sacred of all, the moment of deepest inner concentration. If piety is sincere and expressive of real faith, the contemplation of it has a transforming and awe inspiring effect on those who witness it. Thus also in nature a holy seriousness is to be seen in the fact that natural occurrences are uniformly subject to law [my italics].[15]

A similar distance ahead we have another gateway represented by Retreat. Retreat follows Duration, another static period representing marriage or the enduring union of the sexes. Retreat represents a natural withdrawal. Heaven is retreating. Finally at 34, The Power of the Great, Heaven is underneath Thunder, which seems to represent a large step:

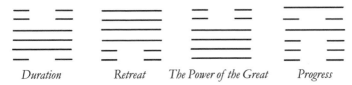

Duration   *Retreat*   *The Power of the Great*   *Progress*

In this hexagram inner worth is said (by Wilhelm) to mount with great force and come to power. The hexagram is followed by Progress.

One could point to the later sequence at 43 to 46, Resoluteness-Coming to Meet-Gathering Together. In Coming to Meet there is only one broken line at the very bottom. One could imagine a soldier crawling under barbed wire. The inferior here is seen as a bold maiden insinuating herself and then seizing power, a real snake in the grass!

Progress through the *I Ching* follows a series of apparent initiation steps which could almost be seen as following stages of life, though the continual return to the beginning suggests rather a cyclic pilgrimage with each cycle hopefully representing an advance in wisdom. Faithful to the cyclic theme, the final hexagram is entitled Before Completion, which immediately follows After Completion.

Wilhelm says at the last hexagram:

> While the preceding hexagram offers an analogy to autumn, which forms the transition from summer to winter, this hexagram presents an analogy to spring, which leads out of winter's stagnation into the fruitful time of summer. With this hopeful outlook the *Book of Changes* comes to its close.

I hope this exercise, following Chungliang Al-Huang's suggestion, may point to a use of the *I Ching* closer to its symbols and leaving aside, for a moment at least, the rather obscure and possibly even misunderstood philosophical passages resonant of Confucius. Except for the sequence including Peace, the passages point to twilight zones of experience where the overwhelming image of Earth is muted to Mountain and Lake, and that of Heaven muted to Thunder and Fire. It is through these transitional symbols that practical progress can take place.

While the diamond and the star symbolise the Earth and Sun, the fundaments of our habitat and the system of which we are a part, it is to the morning star, shining dimly in the twilight, that we look for the hope of its development through our expanding consciousness.

NOTES

1 'Sunset and Evening Star', Tennyson.
2 Advent hymn.
3 'The Hollow Men', T.S. Eliot.
4 C.G Jung, CW 13, 226 and note.
5 I am indebted in much of this discussion to an article *Morning Star* by Jacqueline Keating, submitted to the Jung Institute but unfortunately not published.
6 C.G. Jung, CW 13, 199.
7 'L'Irrémédiable', from *Fleurs du Mal*, Baudelaire.
8 'Hymne à la Beauté', Baudelaire.
9 Robert Pelton, *Trickster in West Africa*, Berkeley Press, 1980.
10 *Ibid.*
11 Richard Wilhelm, *The Secret of the Golden Flower*, RKP, 129n.
12 Chungliang Al-Huang, *Quantum Soup*, Celestial Arts, 96.
13 *Ibid.*, 85.
14 *Ibid.*, 129.
15 Richard Wilhelm, *I Ching*, RKP, 82.

13

In Search of the Soul

Why art thou so heavy, O my soul,
And why art thou so disquieted within me?
Psalm 42

MY INTRODUCTION to the writings of Jung, as with many others I am sure, was his collection of essays entitled *Modern Man in Search of a Soul*. While I was enormously impressed and wished I had not been put off reading Jung at university, I found rather little about the 'soul' as I currently conceived it in terms of my Christian upbringing. There was a lot about the 'psyche' which was the Greek equivalent, but this was used to represent the whole mind of man – as opposed to the body, 'soma'. I could not quite reconcile this. It seemed to me that soul represented for me more than the mind. It included the body, or certain aspects of it, as in the old emergency call SOS, 'save our souls', which obviously related to the body rather more than the spirit, largely regarded as immortal.

Even after many years of reading and analysis it has remained something of a puzzle. Jung's writings are perhaps never quite so ambiguous as on this subject. For the most part references to the soul are replaced by *anima*, or sometimes *animus* when he is specifically looking at women. *Anima* is the Latin translation of *psyche* but since it is conveniently of feminine gender (like psyche) Jung uses it as the inner counterpart of the ego in the male. *Animus* he seems to have invented to use as the opposite in the female. But why does the soul have to be divided by gender?

It is interesting to see what those established authorities Lidell and Scott have to say about the Greek *yuch*. They acknowledge *anima* as the Latin. The first and general meaning is 'breath, especially the sign of life' with a quotation from the death of Aesop. It's secondary meaning is 'the soul of man, as opposed to the body', with a quotation from Homer about a departed soul. So Jung's use of *psyche* is consistent. But the use of *anima* as imparting gender seems spurious.

There are some eight pages devoted to 'soul' in the definitions in CW 6, *Psychological types*, with an interesting footnote by the translators. Jung starts off:

> I have been compelled, in my investigations into the structure of the unconscious, to make a conceptual distinction between *soul* and *psyche*. By psyche I understand the totality of all psychic processes, conscious as well as unconscious. By soul, on the other hand, I understand a clearly demarcated functional complex that can best be described as a 'personality'.

This seems at first splendid. He distinguishes soul from psyche. But why should 'personality' be a mere complex? And is a 'functional complex' different from an 'autonomous complex'? He goes on to speak about different personalities existing in the same person, then about the contrast between the outer personality or outer attitude, the *persona* and the inner personality, or inner attitude, which brings him to define the *anima*. The footnote reads:

> In the German text the word *Anima* is used only twice: here and at the beginning of par. 805. Everywhere else the word used is *Seele* (soul). In this translation *anima* is substituted for 'soul' when it refers specifically to the feminine component in a man, just as in Def. 49 *animus* is sub-substituted for 'soul' when it refers specifically to the masculine component in a woman. 'Soul' is retained only when it refers to the psychic factor common to both sexes. The distinction is not always easy to make, and the reader may prefer to translate *anima/animus* back into 'soul' on occasions when this would help to clarify Jung's argument. For a discussion of this question and the problems involved in translating *Seele* see *Psychology and Alchemy* [CW 12], par. 9 n. 8.

The note in CW 12 adds little, but the passage above is of interest. Speaking of Western man he says:

> If much were in his soul he would speak of it with reverence. But since he does not do so we can only conclude that there is nothing of value in it. Not that this is necessarily always so always and everywhere, but only in people who put nothing into their souls and have 'all God outside.'. (A little more Meister Eckhart would be a very good thing sometimes!)

This is not the place for endless discussion of Jung's works. The soul is essentially *personal* and will continue to have meanings which vary in acceptance widely from individual to individual. For myself, I am content to acknowledge an 'inward face' which holds much of the feminine and which I can regard as the anima and can even paint it. Anima figures in dreams vary from the sublime to the ridiculous and encompass all possible feminine archetypes from the mother to the whore, the queen to the witch. Likewise I can acknowledge and observe that in women the inward face will have much of the masculine, compensating the feminine persona. But it is impossible for me to conceive the anima as defining (or perhaps *confining*) my soul.

Nor can I very happily follow Jung in regarding my soul as a complex. It certainly has emotion, possibly even an emotional centre. But, like the ego with which it has a close association, it is partly conscious and seems able to act not simply autonomously but to an extent through the will. And, at its depth, it seems able to reach closely towards that centre which is the Self.

How else can it be defined except as my individual state of being at any moment? My soul is none other than myself.

Thomas More has written a remarkable book called *Care of the Soul*. More spent part of his life as a monk and I confess it took me a considerable part of the book to overcome my suspicions and even some partly conscious antagonism. Some phrases still 'grate' on my susceptibilities as coming close to 'religiosity'. But at the end I felt rather ashamed of these feelings, since his insight is profound. He goes through a series of what could be called exercises in unwrapping parts of the soul through aspects of pain, neurosis and the 'deadly sins'. His discourses on narcissism and jealousy are wonderful. And (I hope he would agree) he reaches much the same point at the end as I have above.

More brings in at the end the concept of the daemon, and I am reminded of Philip Pullman's wonderful images of children with their daemons attached by a magical cord, the severing of which brings them to a state more awful than death. One can imagine the awfulness of being severed from the unconscious, losing ones soul. More equates daemon with the Latin *animus* but, according to Lidell and Scott, δαιμων, the deity or god, is the Latin *numen*, fate, and can also be 'one's genius', which seems to me the best meaning. More quotes Jung's description of the mistake by the builders at Bollingen giving him an unsuitable stone. Taking the mistake as the work of his Mercurial daemon he transformed it by sculpting it into the famous Bollingen Stone.

But to me, the soul does not equate with the daemon itself. It is rather Philip Pullman's child, with daemon attached which comes closest to my own picture. Although Pullman is not mentioned by More, I feel sure he would agree. After quoting from Ficino and Rilke he says:

> The soul wants to be in touch with that deep place from which life flows, without translating its offerings into familiar concepts ... To live in the presence of the daimonic is to obey inner laws and urgencies.[1]

The daemon can be regarded as a 'guardian angel' if one can accept that an angel need not be a personification of goodness. The picture I have given at Plate 21 is such a symbol of unconscious help and approval.

> When a summer breeze blows through an open window as we sit reading in a rare half hour of quiet, we might recall one of the hundreds of annunciations painters have given us, reminding us that it is the habit of angels to visit in moments of silent reading.[2]

If there is an aspect of the soul which More (and perhaps Jung in some of his passages) may have underrated a little it is the contribution of consciousness. Also I believe it is necessary to take care what one is considering when speaking of consciousness in relation to the soul, for the soul (again in my own conception) is ultimately that 'transparent'[3] individuated being, that amalgam of the conscious and unconscious, which in Jung's terms is making its contribution to universal understanding of the Creator. What I mean here is something simpler. Careful attention to all the senses and the intellect makes an important contribution, as discussed in relation to the life and writing of Iris Murdoch. Here is a passage from *The Sovereignty of Good*:

> It is important too that great art teaches us how real things can be looked at and loved without being seized and used, without being appropriated into the greedy organism of the self. This exercise in *detachment* is difficult and valuable whether the thing contemplated is a human heart or the root of a tree or the vibration of a colour or a sound. Unsentimental contemplation of nature exhibits the same quality of detachment: selfish concerns vanish, nothing exists except the things that are seen.[4]

When More is describing his attitude in listening to his clients, this careful attention stands out, although it is not the subject under discussion.

I believe that, in his book, Thomas More is saying in a different way, a way perhaps more acceptable to some and surely better expressed, what I have been trying to portray through an emphasis on the symbolic: that the symbolic attitude is an attitude that can be regarded as 'care for the soul'.

> To the soul, memory is more important than reason, and love more fulfilling than understanding … Ultimately care of the soul results in an individual 'I' I never would have planned for or maybe even wanted.[5]

NOTES

1 Thomas More, *Care of the Soul*, HarperCollins, 299.
2 *Ibid.*, 303.
3 *Ibid.* Cf. Quotation from Hillman at 261.
4 Iris Murdoch, *The Sovereignty of Good*, RKP, 65.
5 *Ibid.*, 304-5.

14

The Final Sin

MUCH HAS NOW been written about the dangers we are facing if global warming is to proceed unhindered. I have quoted largely from James Lovelock. In *The Revenge of Gaia* he wrote:

I am not a pessimist and have always imagined that good in the end would prevail. When our Astronomer Royal, Sir Martin Rees, now President of the Royal Society, published in 2004 his book *Our Final Century*, he dared to think and write about the end of civilization and the human race. I enjoyed it as a good read, full of wisdom, but took it as no more than a speculation among friends and nothing to lose sleep over.

I was so wrong; it was prescient, for now the evidence coming in from the watchers around the world brings news of an imminent shift in our climate towards one that could easily be described as Hell: so hot, so deadly that only a handful of the teeming billions now alive will survive. We have made this appalling mess of the planet and mostly with rampant liberal good intentions. Even now, when the bell has. started tolling to mark our ending, we still talk of sustainable development and renewable energy as if these feeble offerings would be accepted by Gaia as an appropriate and affordable sacrifice.[1]

Hardly the words of an optimist. But no authors I have read are as persuasive as Primack and Abrams:

Saint Augustine enunciated the Christian doctrine: 'The deliberate sin of the first man is the cause of *original sin*.' Whether you believe that or not, failing to protect our species and destroying the promise of the only intelligent life that may exist would surely be a *final sin*. It could extinguish the Sovereign Eye of the universe.[2]

The authors took the symbol of the sovereign eye from Shakespeare's Sonnet 33 as symbolising intelligent life within the cosmos. They also

152

point out that global warming is not the only, nor perhaps even the most immediate danger to the human species. There is the rapid expansion of the population, an expansion which is exponential.

> The world is at a turning point. Not the turning point of this election cycle, not even the turning point of a lifetime, but a turning point that can only happen once in the evolution of our planet. Some may dismiss this as a ridiculous exaggeration, since it is so unlikely that such a momentous turning point would occur in *our* short lifetimes. Unlikely or not, it's here. If we take our cosmic role seriously and let our largest selves find the sanest way across the mountains, we can come down the other side having created a stable and wise long-term civilization that will allow our descendants to benefit from the amazingly benign conditions of our beautiful planet. If we don't, they may curse us forever.[3]

The curve they show of population growth over the last 2000 years is dramatic. During the twentieth century the human population quadrupled. It is likely to double again within a generation and, as Primack and Abrams point out, 'no one seriously proposes that the earth can sustain another population doubling'.[4] They draw an analogy from Goethe's 'The Sorcerer's Apprentice', whose broomstick kept multiplying as it broke, and go on to say:

> All exponential growth in a finite environment follows similar patterns. Exponential growth reaches a physical limit and ends – perhaps smoothly, perhaps abruptly. When growth overshoots a limiting resource, the population collapses. The exponential growth in human population and resource use must end one way or another in the current generation.[5]

This is quite a shattering statement, coming on top of the crisis in global warming. One wonders which catastrophe will hit us first. The first tragedies of global warming are already upon us: the tsunamis and hurricanes, the rise in sea level which will soon hit low-lying countries. The shrinking of the ice cap is not only leaving polar bears stranded, but is increasing the rate of warming as reflection from the snow reduces. And in fact resource use is inflating faster than population. Population rise is conspiring with aridity and warfare and economic expectations in poorer regions of the world to cause mass migration which will certainly accelerate.

The pattern of cultural or tribal distribution in the more developed countries, due to random immigration, is already causing stress and undermining generations of established mores which have hitherto stabilised nation states.

Amid this scene of chaotic change Primack and Abrams glimpse a small hope. We cannot expect the sorcerer to arrive and halt the chaos. However the population rise need not stop altogether, if one can take as an example the history of the universe. Within the first fraction of a second our present universe expanded at an inflationary rate, but then did not stop abruptly but 'slowed to a crawl'.

> Only then did it enter its most creative phase during which it produced galaxies, stars, planets and life. The fundamental character of the universe has been to grow in complexity.[6]

They suggest that we need to start seeing the coming changes as opportunities not to *acquire* more but to *become* more. Every sorcerer, they say in hope, began as an apprentice. I am reminded of the biblical text 'Fear of the Lord is the beginning of wisdom' (Proverbs 9.10). For the Lord read, perhaps, Gaia.

Lovelock also touches on the population explosion:

> The root of our problems with the environment comes from a lack of con-
> straint on the growth of population ... If we can overcome the threat of
> deadly climate change ... our next task will be to ensure that our numbers
> are always commensurate with our and Gaia's capacity to nourish them ...
> At first this may seem a difficult, unpalatable, even hopeless task, but events
> in the last century suggest that it might be easier than we think. Thus in
> prosperous societies, where women are given a fair chance to develop their
> potential they choose voluntarily to be less fecund ... It is only a small step
> ... but it is a seed of optimism from which other voluntary controls could
> grow and surely better than the cold concept of eugenics that withered in
> its own amorality.
>
> The regulation of fecundity is a part of population control, but the
> regulation of the death rate is also important. Here, too, people in affluent
> societies are choosing voluntarily seemly ways to die ... Now that the Earth
> is in imminent danger ... it seems amoral to strive ostentatiously to extend
> our personal lifespan beyond its logical limit of about one hundred years ...
> we have to make our own constraints on growth and make them strong and
> make them now ...

In the end, as always, Gaia will do the culling and eliminate those who break her rules.

Even one hundred years can be terrible for those caught up in the end game of institutional life, often twenty years or more crippled by arthritis and befuddled by Alzheimer's disease. Even in the throes of a hopeless ordeal like motor neurone disease the current dictates of the dogmatic religious movements prohibit a kindly end permitted to creatures lucky enough not to possess the questionable gift of consciousness. The end of life as well as its procreation is in great need of reassessment. And it would seem that Lovelock may have been hasty in believing that 'in prosperous societies, where women are given a fair chance to develop their potential they choose voluntarily to be less fecund.' There is, on the contrary, a current trend towards large families among the affluent.

Little enough attention has been paid to the efforts of Gaia to maintain a balance by culling. I am not clear what Lovelock had in mind, but suspect he was referring to global warming. However in all divisions of life there is a natural striving against excess by the provision of some mechanism of negative feedback, that great mechanism brought into play by the presence of opposites, and the interventions of science and husbandry are very largely directed to reducing this effect. It is surely no accident that every advance in antibiotics is met with a new, resistant mutation, the first and most obvious being HIV, those more current including MRSA, *Clostridium difficile* and avian influenza. Some such disease may prove to be the trigger of a saving butterfly effect. We cannot cease striving to eliminate such diseases but, if we continue to do so, we must concurrently seek to deny Gaia her revenge by reducing global population in other ways, and urgently. Voluntary euthanasia would seem to be an essential first step. Luckily the medical profession no longer seems to be the main opponent and in some countries assisted suicide is tolerated. As so often in history it is the 'religious' who need to balance principle with the quality of mercy.

Looking around at the terrible unwisdom of the human race, one wonders where and how the necessary renewal could possibly start. With small acts of imagination, perhaps, placing hope in invoking a more tolerable butterfly effect. Although reaching this hope myself, as I write, I note that Primack and Abrams came to this conclusion also.[7] The following passage is very reminiscent of Jung:

The ancient Egyptians saw themselves upholding the cosmos itself by upholding Order, Harmony, and Truth through their rituals; ... We matter to the universe as we know it, because the universe as we know it dies without us. We uphold the new universe – but only if we too, like the ancients, consciously do uphold it in our thoughts and actions.[8]

Jung's conversations with the native Americans also shows the import-ance of symbolic ritual which we may need to recover:

The religious leader of the Taos pueblo, also known as the Loco Tenente Gobernador, once said to me: 'The Americans should stop meddling with our religion, for when it dies and we can no longer help the sun our Father to cross the sky, the Americans and the whole world will learn something in ten years' time for then the sun won't rise any more.' In other words, night will fall, the light of consciousness is extinguished, and the dark sea of the unconscious breaks in.

Whether primitive or not, mankind always stands on the brink of actions it performs itself but does not control.[9]

While Jung interpreted Gobernado's words as portending domination by unconscious forces, they could equally portend the end of a bene-volent nature. Primack and Abrams' vocation is to move humanity collectively towards a new myth. 'We need,' they say,

collectively, to become the kind of people capable of using science to uphold a globally inclusive, long-lived civilization. How do we get there? By facing up to the imperative that we have to face up to. As H.G. Wells put it, 'Human history becomes more and more a race between education and catastrophe.'[10]

I see this as compatible with Jung's vocation to move individuals towards a higher state of consciousness. In my view the two are com-plementary. And both rely on the need to seek and acquire a symbolic attitude. Both are urgently needed. Indeed the need is now so urgent that one can only hope that they can somehow progress together.

What hope can be held out? Primack and Abrams take a meaningful view which holds a germ of optimism.

The only place beings with a consciousness like ours can ever feel ourselves *belonging* to the universe is at its centre. But the longing to be central is not what makes us central: the structure of the universe makes us central.[11]

They contrast this with what they term the existentialist view that 'in the expanding universe "human" is a small, even pathetic entity', which they proceed to demolish:

The difference between the existential view and the meaningful view is not just 'The glass is half empty' versus 'The glass is half full,' because everyone knows what a glass looks like and what the alternatives mean. But if we resign ourselves to being some minor trash in the universe, we will *never see* what the universe looks like, because that can only be seen with the mind's eye, and the mind's eye works from metaphors that are inaccessible to people who hold the humans-as-trash assumption.

The meaningful universe encompasses the existential in the sense that the meaningful can understand the existential, but the existential cannot see the meaningful.[12]

The symbols of the new age and the rebirth of a search for the spiritual may have arrived at the critical time. Resistance to this comes, to my mind, not so much from the establishments: political and business, conditioned to sway with change, as from the dogmatic religions and sects who seem to see in it a challenge to their neo-tribal coteries. From this perspective the most depressing example in the last half century must be Palestine. The most encouraging area has been South Africa, blessed with the charismatic figures of Nelson Mandela and Desmond Tutu. In the present or near future it seems, paradoxically, to lie in Northern Ireland. For so many years divided by religious and cultural – essentially tribal – extremes, a tentative *rapprochement* and synergy seems to be taking place. Although presently more practical than spiritual, it has the potential for a symbol of reconciliation. It may yet prove to be a butterfly of change.

Brooding on the ugliness of our present civilisation I am reminded of Alan Watts' book, perhaps his greatest contribution, *The Wisdom of Insecurity*. Watts' message is vividly illustrated by Chungliang Al-Huang:

Alan Watts once wanted to use the title 'Shangri-la on the San Andreas Fault' for a seminar he was giving at an institute on California's Big Sur coastline. However, the management decided it was unwise to advertise their proximity to the famous fault and changed the title to 'Shangri-la at the Edge of the Sea.' Exit paradox, enter non sequitur.

Alan's point, of course, was that for true peace of mind we must acknowledge whatever fault we live upon, whatever time bomb ticks in our closet, and enjoy our Shangri-la nonetheless. It is not the absence of the problem; it is how one lives in its presence that matters.[13]

I once attended a seminar by Baroness Vera von der Heydt which revolved around the concept of uncertainty. Uncertainty and insecurity are closely related, and related also to unknowing, as expressed by the author of *The Cloud of Unknowing*. They are part of the human condition, never more so, and need acceptance, and more, embracing. For certainty and security, and indeed full understanding are all sterile, indicative only of death, the *t'ai chi* frozen into immobility.

As I write, the global establishments, so long dragging their heels, are at long last picking up the warnings broadcast by the advance guard of environmental scientists and societies, albeit with predictable reluctance and lethargy. This has been advanced by scientists I have described, not ashamed to acknowledge a spiritual quality in their work and writing, and a sense of vocation. 'Holism', although disguised under other names, is not the anathema it was. Perhaps most noteworthy is the symbolic attitude which seems to pervade their contribution.

The academic world is picking up new attitudes. There is a proliferation, now, of faculties with new areas of study broad enough to embrace psychologists like C.G. Jung who was, in my time as a student, held in profound disregard by the medical, philosophic and theological professions as a pagan mystic. One of the great advances has been the founding of The Scientific and Medical Network to which many of the authors I have mentioned, including John Clarke, David Tracey, Margaret Arden, Joel Primack and Nancy Abrams, have contributed. Its founders seem to have been inspired by a symbolic attitude. I hope this will prove a chrysalis for another butterfly. John Clarke writes in his book *Jung and Eastern Thought*:

> In his analysis of the relevance of the new physics to the principle of synchronicity he noted the importance of the idea of pattern and interconnectedness, rather than of interaction between discrete particles as in classical physics. From this quarter there was emerging, he surmised, a new unitary conception of being, one which not only breaks down absolute barriers between entities and between space and time, but between observer and

observed, between subject and object, between mind and matter. It points, in other words, to what would nowadays be described as a *holistic* concept of being.

Now it was precisely this quality of the *I Ching* that he seized on ...'[14]

To receive the message of the *I Ching* requires above all a symbolic attitude. One could almost say that, with this attitude, the *I Ching* is no longer a necessary vehicle. The message is ever present: *Turn but a stone, and start a wing.*

I will let one of my favourite people, the Quaker E.F. Schumacher, have the last word:

... industrial society, unless radically reformed, must come to a bad end. Now that it has – adopted cumulative growth as its principal aim, its end cannot be far off. But that does not mean that it will have failed in its purpose from the point of view of the Gospel. Out of the tremendous examination set by this monstrous development many single individuals will emerge triumphant; uncorrupted and hence incorruptible. This is all that really matters ...

Why should industrial society fail? Why should the spiritual evils it produces lead to worldly failure? From a severely practical point of view, I should say this:

1. It has disrupted, and continues to disrupt, certain organic relationships in such a manner that world population is growing, apparently irresistibly, beyond the means of subsistence.

2. It is disrupting certain other organic relationships in such a manner as to threaten those means of subsistence themselves, spreading poison, adulterating food, etc.

3. It is. rapidly depleting the earth's nonrenewable stocks of scarce mineral resources – mainly fuels and metals.

4. It is degrading the moral and intellectual qualities of man while further developing a highly complicated way of life the smooth continuance of which requires ever-increasing moral and intellectual qualities.

5. It breeds violence – a violence against nature which at any moment can turn into violence against one's fellow men, when there are weapons around which make nonviolence a condition of survival.

It is no longer possible to believe that any political or economic reform, or scientific advance, or technological progress could solve the life-and-death problems of industrial society. They lie too deep, in the heart and soul of

160 *The Diamond and the Star*

every one of us. It is there that the main work of reform has to be done –
secretly, unobtrusively. I think we must study nonviolence deep down in our
own hearts. It may or may not be right to 'ban the bomb.' It is more important to overcome the roots out of which the bomb has grown. I think these
roots are a violent attitude to God's handiwork instead of a reverent one.
The unsurpassable ugliness of industrial society – the mother of the bomb
– is a sure sign of its violence.[15]

While I was sitting in my garden brooding on whether to embark on
this project I happened to look up at the sky and saw a cloud formation
shaped in what seemed to be the great wing of an angel (Plate 21). It
reminded me of Iris Murdoch, about whom I had been reading; of her
vision of the kestrel when looking out of her window.

In a moment everything is altered. The brooding self with its hurt vanity
has disappeared. There is nothing now but the kestrel ... 'Not how the world
is, but that it is, is the mystical.'[16]

I no longer felt alone and powerless. My guardian angel would be my
companion.

Returning to the divine marriage of the diamond and the star, perhaps
the star may express the mystery of the great cosmos, and the diamond
the seed of life buried in our planet, their offspring the divine child of
human consciousness, the Word made flesh, born anew at the turn
of every year. May it yet hold our saving grace.

NOTES

1 James Lovelock, *The Revenge of Gaia*, Penguin, 147.
2 Joel Primack and Nancy Ellen Abrams, *The View from the Centre of the Universe*, Fourth
 Estate, 294.
3 *Ibid.*, 261.
4 *Ibid.*, 254.
5 *Ibid.*, 255.
6 *Ibid.*, 256.

7 *Ibid.*, 260.
8 *Ibid.*, 295.
9 C.G. Jung, CW 9i, 48.
10 Primack, *ibid.*, 296.
11 *Ibid.*, 272.
12 *Ibid.*, 275.
13 Chungliang Al-Huang, *Quantum Soup*, Celestial Arts, 59.
14 J.J. Clarke, *Jung and Eastern Thought*, Routledge, 98.
15 E.F. Schumacher, *Good Work*, Abacus, 35, 36.
16 Iris Murdoch, *The Sovereignty of the Good*, RKP, 84, 85.

Appendix I
'Participation Mystique' and Mysticism

THE ATTENTION given by Jung to 'participation mystique' in the foreword to *The Secret of the Golden Flower* requires for its clearer understanding some exploration of this important concept. It involves an understanding of Jung's view of the place and significance of encounters with the unconscious and their differences within primitive cultures on the one hand and modern or Western culture on the other. In the former they are the norm and therefore generally benign, but in the latter they can appear distorted and potentially dangerous. However they are also responsible for the important phenomenon of mysticism.

Definitions

There are over twenty references to 'participation mystique' given in the General Index to Jung's *Collected Works*. All of them are under the heading Levy-Bruhl, to whom the expression is due, from his book *How Natives Think*. A start may be made in the definition given in Vol. 6 (*Psychological Types*) at paragraph 781 (the italics refer to earlier definitions):

PARTICIPATION MYSTIQUE is a term derived from Levy-Bruhl. It denotes a peculiar kind of psychological connection with objects, and consists in the fact that the subject cannot clearly distinguish himself from the object but is bound to it by a direct relationship which amounts to partial *identity*. This identity results from an *a priori* oneness of subject and object. *Participation mystique* is a vestige of this primitive condition. It does not apply to the whole subject-object relationship but only to certain cases where this peculiar tie occurs. It is a phenomenon that is best observed among primitives, though it is found very frequently among civilized peoples, if not with the same incidence and intensity. Among civilized peoples it usually occurs between persons, seldom between a person and a thing. In the first case it is a transference relationship, in which the object (as a rule) obtains a sort of magical

163

– i.e. absolute – … influence over the subject. In the second case there is a similar influence on the part of the thing, or else an *identification* with a thing or the idea of a thing.

The reference back to the definition of 'identity' is important:

IDENTITY. I use the term identity to denote a psychological conformity. It is always an unconscious phenomenon since a conscious conformity would necessarily involve a consciousness of two dissimilar things, and, consequently, a separation of subject and object, in which case the identity would already have been abolished. Psychological identity presupposes that it is unconscious. It is a characteristic of the primitive mentality and the real foundation of *participation mystique*, which is nothing but a relic of the original non-differentiation of subject and object, and hence of the primordial unconscious state. It is also a characteristic of the mental state of early infancy, and, finally, of the unconscious of the civilized adult, which, in so far as it has not become a content of consciousness, remains in a permanent state of identity with objects. Identity with the parents provides the basis for subsequent identification with them; on it also depends the possibility of *projection* and *introjection*.

Identity is primarily an unconscious conformity with objects. It is not an equation, but an *a priori* likeness which was never the object of consciousness. Identity is responsible for the naïve assumption that the psychology of one man is like that of another, that the same motives occur everywhere, that what is agreeable to me must obviously be pleasurable for others, that what I find immoral must also be immoral for them, and so on. It is also responsible for the almost universal desire to connect in others what most needs connecting in oneself. Identity, too, forms the basis of suggestion and psychic infection. Identity is particularly evident in pathological cases, for instance in paranoic ideas of reference, where one's own subjective contents are taken for granted in others. But identity also makes possible a consciously *collective* social *attitude* which found its highest expression in the Christian ideal of brotherly love.

'Primitive' Cultures

The primitive Bushmen, whose 'participation mystique' is so marvelled at by Sir Laurens van der Post (see Chapter 8) and whose visions are achieved through dancing, are elsewhere described by him in terms which one can only regard as highly individuated:

Did it really matter whether the end came from the crab within or from the hyena without? We will have courage to meet it and give meaning to the character of our dying provided we, like these humble, wrinkled Bushmen, have not set a part of ourselves above wholeness of life.[1]

There is a remarkable agreement between this passage and Jung's commentary on the meditative exercise in *The Secret of the Golden Flower*:

In this remarkable experience I see a phenomenon resulting from the detachment of consciousness, through which the subjective 'I live' becomes the objective 'it lives me'. This state is felt to be higher than the earlier one; it is really as though it were a sort of release from the compulsion and impossible responsibility which are the inevitable results of *participation mystique*.[2]

The responsibility is one which must be peculiar to Western man. To the primitive, at least, the gods and demons are the responsible ones. It is as though Jung in this passage grants to the Chinese alchemists a higher state of consciousness than the primitive Africans or native Americans (below), in spite of his attribution to them of a predominantly unconscious attitude.

Jung's adopted concept of 'participation mystique' is not easy for a modern Westerner to assimilate. Even Sir Laurens van der Post only understood its significance under the spell of the Bushmen. Jung describes a similar experience in the story in Vol. 8, *The Structure and Dynamics of the Psyche*, of the Africans on Mount Elgin:

... an old chief began to explain. 'It is so,' he said. 'When the sun is up there it is not God, but when it rises then it is God' ... Sunrise and his own feeling of deliverance are for him the same divine experience ... For him night means snakes and the cold breath of spirits, whereas morning means the birth of a beautiful god.[3]

The above is on a plane with van der Post's Bushmen. However, in the following illustration of the story of Hiawatha killing his first roebuck, the conscious attitude expressed by Jung is arguably more reminiscent of the Chinese alchemists:

... the roebuck was no ordinary animal, but a magic one with an unconscious (i.e. symbolic) significance ... In killing his first roebuck, therefore, Hiawatha was killing the symbolic representative of the unconscious, i.e. his own *participation mystique* with animal nature, and from that comes his giant strength.[4]

How far *Hiawatha* is a projection of Longfellow's is perhaps unknowable. But the story is also reminiscent of other stories from van der Post's Eranos Lecture. The tale of the eland 'provoked the social and family aspects of the Bushman's life into consciousness'. The slaying of the lion is a tale of individuation. The tale is a long one in the original words and is necessarily abbreviated:

A young man goes out to hunt a lion and falls asleep. A thirsty lion, on his way to drink, attacks the sleeping man and, believing him dead, fixes him in the fork of a thorn tree. But the man is not dead. Feeling his head caught on a spike in the tree, 'tears of fear and anguish come running down his cheeks' and he turns his head. The lion notices the movement, returns and, seeing the tears, starts licking them away. A transformation then takes place and the lion 'is linked for good or ill to the hunter and no one else'. The lion disappears to drink and the young man seizes the opportunity to flee to his community, running zigzag because he knows the lion will come after him. He begs the community to wrap him for protection in antelope skins. They do so because, 'It was a young man their hearts loved very much.' When the lion arrives the community sets out to kill him but the lion appears indestructible. It says only, 'I have come for the young man whose tears I have licked and whom I must find.' The community decide, in spite of the cries of the man's mother, that they cannot have the lion in their midst; they remove the skins and throw him to the lion. The lion approaches and says, 'I have found the young man whose tears I have licked, and now you can kill me and I'll die.' At that moment the community is enabled to kill the lion.

Van der Post interprets this in the following words:

> There must be between you and your royal, natural self (the lion is the king of the beasts) an acknowledgement, a coming to terms with nature which you cannot expect the community to make for you. In this matter you cannot hide indefinitely behind the attitude, the conventions and morals of the community: that is merely being wrapped up in a protective covering of the skins of the community. You must not fall asleep on the way to the water of life. You have to make your own terms with the animal within you. You have to stand on your own two feet and so become an individual and courageous man.[5]

What can we learn from these mythic stories? In themselves they are tales of young men growing up, facing a rite of passage into manhood. But they also acknowledge in quite primitive societies an ability to differentiate, even to individuate. They do more than merely demonstrate an individual consciousness in the sense of individuality:

> The psychological individual, or his *individuality* ... exists only so far as there exists a conscious distinction from other individuals ... A conscious process of *differentiation* or *individuation* is needed to bring the individuality to consciousness. i.e. to raise it out of the state of identity with the object.

> Individuation, therefore, is a process of *differentiation* having for its goal the development of the individual personality ... Individuation is closely connected with the *transcendent function* since this function creates individual lines of development which could never be reached by keeping to the path prescribed by collective norms.[6] [Italics refer to other definitions.]

That individuation is to be found in primitive societies should not come as a surprise when one considers the great figures of antiquity, but it must be borne in mind that individuation from an unconscious norm will be different from individuation from a conscious norm. Also one must be very careful where one throws the attribute 'primitive'. Many ancient cultures were of course as civilised – or more so – than that of the modern West. This could certainly be said of the ancient Chinese whom Jung regarded as sophisticated psychologists.[7] That their standpoint or norm can be regarded as 'unconscious' seems to me less a matter of psychological development than of culture. Unlike the West, their written language was (and remains) not linear but pictorial, based upon images or ideograms. According to Alan Watts:

> Reading Chinese is fundamentally what communications technicians call 'pattern recognition' – a function of the mind which is, as yet, only rudimentarily mastered by the computer because it is a nonlinear function.

> Alphabetic writing is a representation of sound, whereas the ideogram represents vision, and furthermore represents the world directly – not being a sign for a sound which is the name of a thing. As for names, the sound 'bird' has nothing in it that reminds one of a bird.

> Just as Chinese writing is at least one step closer to nature than ours, so the ancient philosophy of the Tao is of a skillful and intelligent following of

the course, current and grain of natural phenomena – seeing human life as an integral feature of the world process, and not as something alien and opposed to it.[8]

This difference in norm – culture, outlook, tradition, *weltanschauung*, whichever may be most apt, is huge, and one wonders how far afield it is spread. The difference is also blurring as non-Western societies are developing a Western culture; furthermore the rapid industrial development of China is forcing Westerners, in their turn, to assimilate Chinese.

The Chinese appear unique in maintaining ideogrammatic writing and even speech. Although the Japanese employ Kanji characters, the spoken language is syllabic, and it is necessary to speak Japanese in order to read it. It is not necessary to speak Chinese to read it: it has been found that second grade children who were backward in reading could easily be taught to read Chinese, and could construct simple sentences within four weeks.[9]

The Indian languages, on the other hand, including the ancient Sanskrit, Arabic and languages developed from it such as Malay and Swahili, are all linear, alphabetic languages. So far as I know, this extends to African tribal languages. While psychological development of these cultures may be expected to pursue that of the West, the difference between the Hindu and Buddhist traditions, on the one hand, and the monotheistic tradition of Islam cannot be ignored.

'Western' Culture

There is a poignant reference to Meister Eckhart in CW 6, where Jung quotes from a mystical experience of Eckhart:

> … in this breakthrough I receive what God and I have in common. I am what I was, I neither increase nor diminish, for I am the unmoved mover that moves all things. Here God can find no more place in man, for man by his emptiness has won back that which he eternally was and ever shall remain.[10]

Jung regards this experience as a return to a primitive consciousness:

> But when the 'breakthrough' abolishes this separation [ego and subject] by cutting off the ego from the world, and the ego again becomes identical with the unconscious *dynamis*, God disappears as an object and dwindles into a subject which is no longer distinguishable from the ego. In other words the

ego, as a late product of differentiation is reunited with the dynamic All-oneness (the *participation mystique* of primitives).

This is a paradox. Is Jung advocating a return to primitive unconsciousness? Meister Eckhart is actually a favourite figure, most significantly in describing the relativism of the God image:

Both on account of his psychological perspicacity and his deep religious feeling and thought, Meister Eckhart was the most brilliant exponent of that critical movement within the Church ...

We now come to the main theme of this chapter – the relativity of the symbol. The 'relativity of God,' as I understand it, denotes a point of view that does not conceive of God as 'absolute,' i.e., wholly 'cut off' from man and existing outside and beyond all human conditions, but as in a certain sense de-pendent on him; it also implies a reciprocal and essential relation between man and God, whereby man can be understood as a function of God, and God as a psychological function of man.[11]

As Meister Eckhart puts it:

The soul is not blissful because she is in God, she is blissful because God is in her. Rely upon it, God himself is blissful in the soul.[12]

This almost equivalent reciprocity may have been an inspiring factor for Jung's concept of the evolution of the God image through man's developing consciousness portrayed in *Answer to Job* and *Memories, Dreams, Reflections*. Jung goes on to explain the mystical experience in these terms:

Psychologically, this means that when the libido invested in God, i.e., the surplus value that has been projected, is recognized as a projection, the object loses its overpowering significance, and the surplus value consequently accrues to the individual, giving rise to a feeling of intense vitality, a new potential. God, life at its most intense, then resides in the soul, in the unconscious. But this does not mean that God has become completely unconscious in the sense that all idea of him vanishes from consciousness. It is as though the supreme value were shifted elsewhere, so that it is now found inside and not outside. Objects are no longer autonomous factors, but God has become an autonomous psychic complex. An autonomous complex, however, is always only partially conscious, since it is associated with the ego only in limited degree, and never to such an extent that the ego could wholly comprehend it, in which case it would no longer be autonomous. Henceforth the determining factor is no longer the overvalued object, but the unconscious. The determining influences are now felt as coming from within oneself, and this feeling produces a oneness of being, a relation between conscious and unconscious, in which of course the unconscious predominates.[13]

Those who regard Jung as a mystic are not wrong, although those who dismiss him as '*merely* a mystic' show a thoughtless bias. The truth is that he was fascinated by the phenomenology of mysticism and its context within the larger phenomenon of the individuation process. A mystical experience is short-lived. I am reminded of the hexagram for 'grace' in the *I Ching* discussed earlier. It can, in some, be dismissed as a passing fancy. But for most it is a defining moment, never to be forgotten. The course of individuation follows a lifetime of experiences. It is never finished. It follows a spiral course of progression and regression: a continuous *reculer pour mieux sauter*. But a mystical experience, once encountered, is a platform. It is usually repeated, sometimes with years in between, and in the real mystics it is a near – though never wholly – normal state.

What is important is that such an experience must never overwhelm. That way lies gross inflation and the madness of infallibility: a Hitlerian possession by Wotan. Which is why the experience must always be partially conscious. And to the extent that we are responsible for the evolution of God, it is through an evolving consciousness.

I would like to finish with Iris Murdoch:

Not how the world is, but that it is, is the mystical.[14]

NOTES

1 Sir Laurens van der Post, *The Lost World of the Kalahari*, Penguin, 252.
2 Richard Wilhelm, *The Secret of the Golden Flower*, RKP, 132.
3 C.G. Jung, CW 8, 411.
4 C.G. Jung, CW 5 (*Symbols of Transformation*), 504.
5 Sir Laurens van der Post, *Eranos Lectures*, 5, Spring Publications, 37.
6 *Ibid.*, CW 6, 755-60.

7 Richard Wilhelm, *ibid.*, 129.
8 Alan Watts, *Tao: The Watercourse Way*, Pelican, 13-16.
9 Alan Watts (*supra*) page 11 quotes Rozin, Poritsky and Scotsky, *Science*, March 26, 1971, 1264-7.
10 C.G. Jung., CW 6, 255.
11 *Ibid.*, 243.
12 *Ibid.*, 246.
13 *Ibid.*, 248.
14 Iris Murdoch, *The Sovereignty of Good*, RKP, 85.

Appendix II
Excerpts from the Original Essay

Foreword

THIS STUDY is, in one sense, the culmination of an odyssey I made at the C.G. Jung Institute in Küsnacht, near Zurich, in 1987 and 1988, and of an analysis which I undertook there. In another sense, it is a beginning. Perhaps it was my background in Chemistry, and my experience of analysing and re-synthesising the inventions of others, to crystallise in my mind the essence of their novelty, so that they could be protected by the 'temenos' of the law and my fascination with the strange chemistry of carbon, that led me to experience the numinosity of the diamond within the symbology of the Self.

Perhaps, also, it was my unexpected fascination with the Tao, which led me to attempt to re-write the *Tao Te Ching* of Lao Tsu in my own words, and to experience a week's course in it Austrian Alps under *T'ai Chi* Master Chungliang Al-Huang and my growing understanding and appreciation of the *I Ching*. For it was the Chinese alchemists who intuited the three-dimensional quaternary structure of the diamond, and used it as a symbol of individuation 3000 years before Dorothy Hodgkins developed the science of crystallography.

But I cannot explain the star; it simply arrived. Or perhaps it was always there, waiting for the diamond.

Drawing the Mandalas

Our father Adam sat under the Tree and scratched
with a stick in the mould;
And the first rude sketch that the world has seen
was joy to his mighty heart.

Till the Devil whispered behind the leaves,
'Its pretty, but is it Art?'[1]

FOR ITS CLEAR understanding this section should be read in conjunction with Plate 1, showing the two mandalas.

The Diamond Mandala

The Birth

I HAD BEEN trying to prepare a structure for a tetrahedron in the form of an atomic skeleton from plasticine and toothpicks, seeking to get a view of the three-dimensional form from different angles. This followed a question in my mind why Jung, with his emphasis on four-sided figures, had never really looked at the tetrahedron, the crystal form of the diamond. Unfortunately the materials were not strong enough and I had to defer this exercise until I could get hold of proper teaching models. I had read that the diamond did contain cubic symmetry, with atoms at the centres of the sides of a cube. I tried to draw this on paper, but could not find a way of showing this. Tetrahedrons are also extremely difficult to draw, from any angle.

Strangely, my next impetus towards the mandala came from dreams. The dreams seemed to be preparing me, at least unconsciously, for a crisis point in my life. I had had two successive dreams involving a pair of keys. In both dreams a third key was lost; in one dream I had lost the key to my home; in the other an anima figure had lost the key to her office.[2] I searched some literature for references to keys, and in Jung's seminar on *Dream Analysis* found a reference to the *ankh* as a the 'Key of Life'. This was given as a form of cross. On the next page my eye was drawn to a Teutonic cross, similar to the Maltese cross, the Templars' cross or cross of St. John, but formed of four equilateral triangles. It came to me that these were the separated sides of a tetrahedron joined at a point or axis. I saw that it also resembled a four-leaf clover, that rare find which is an emblem of luck.

I did some doodling on paper and saw that three of the 'sides' could be pivoted up or down from the axis to form three sides of a tetrahedron, thereby defining the fourth, but that this could only be filled by removing the last side from the cross and placing it in position. At that point some unconscious impulse made me draw the three-leaf clover or shamrock, the emblem of Ireland, associated with the leprechaun, the 'trickster'. It was quite difficult to get this symmetrical, and, when at last I had it to my satisfaction, I saw that the sides seemed to be continuous straight lines which formed diameters of a circle circumscribing the figure. I realised that, if the triangles were again pivoted downwards, the resulting tetrahedron would lie on the bottom of the corresponding sphere. So I drew in the triangle and the circumscribing circle. It was difficult to place the triangle exactly centrally, but I became excited when I saw a resemblance to the Shri Yantra.[3] I felt a strong urge to draw the figure with mathematical precision to avoid erroneous conclusions and to discover any further possible associations with the diamond.

The Tools

When I embarked upon this step, I was immediately frustrated by a total lack of instruments, such as compasses, ruler, or protractor.[4] After a pause, in the Zen tradition, I searched around to see what I could find and immediately noticed a round black plastic coffee cup, exactly the size I wanted, and a transparent cassette holder which I could take to pieces to give exactly the right length of line and a precise right angle. The only other instrument I needed later was a knife to nick the plastic with the precise length of the radius.

When I started work, I found that I fell into a meditative rhythm; each step completed produced a great satisfaction and I found myself antici-pating with pleasure the next problem to be posed and the solution which I knew would arrive from the unconscious.

Dividing the Circle

I first drew round the cup to produce the circle. The immediate problem was to draw the horizontal diameter exactly. I started by drawing as best I could a vertical tangent to the left, perpendicular to the top of the sheet, and followed this by another on the right. I then carefully marked the precise points of tangent with a pin and drew the diameter. The next problem was to find the centre.

The Square

At that point I found myself looking at the two tangents with an urge to join the ends into a circumscribing square, and realised that the corners of the square were exactly what were needed to fix the centre. I found that the two tangents were not precise enough for my liking and redrew them exactly perpendicular to the diameter. I then drew the upper and lower tangents precisely at right angles. The resulting embryo mandala was very satisfying.

Finding the Centre

Locating and marking the centre was no real problem using the corners, but this location was so critical that I decided that it was necessary to use a sharp knife to mark the paper. This use of the knife in turn led to the idea of nicking the plastic edge of the cassette holder with the exact length of the radius. I was able to double check the position of the centre which I found remarkably exact.

The Visible Sides

I then saw that my 'radius measure' was exactly what I needed to mark the apices of the triangles, since each side was of radial length. Using the measure, I was able to mark points precisely, again with the knife, and with great satisfaction drew in the shamrock figure. I drew it with feminine triangles, which I felt compelled to do.[5]

The Error

The final problem was how to locate and draw the central triangle forming the base of the tetrahedron and at this point I made an error. I could not immediately see how to mark the location of the apices, nor indeed where the centre of the triangle should be. Looking at the rough sketch, it appeared that the top of the centre triangle should be half way along the height of the top triangle. I studied the figure to see how I could bisect the triangles and saw that a line drawn between alternate apices would pass through the intermediate radius at the half-way point. But when I had drawn some lines, I saw that what I had done was in fact to draw a rectangle which was not a square. It took a while to recognise what had happened. It is still somewhat perplexing, but it resulted from the facts that I had not properly differentiated 'triangles' from 'gaps' and that the original sketch was quite misleading: the top of the centre triangle was certainly not half way up the top triangle.

The Hidden Base

I mention this error in some detail, because I think that the enforced meditation that followed resulted in some critical insights. I found that the lines I had drawn gave six other points which I had not noticed before, which formed the inner apices of a six-pointed star of which the points were given by the outer triangles. Three of these points were at what I immediately recognised as the centres of the three drawn triangles, and furthermore they were obviously half way along the radius of the circle. I could then see that the remaining three 'crossing points' defined my inner triangle. With immense satisfaction I finally drew this in.

The Unfolding

When I came to paint the mandala I chose to make the three 'upper' triangles gold, and the 'base' grey. This brought to mind the unconscious inferior function, partly lit by the other functions, and partly obscuring them. Centring the tetrahedron within the sphere would mean raising the dark base and tilting down the golden sides, like folding an umbrella. But the point of pivot would rise along with the base, until the centre of the fully formed tetrahedron coincided with the centre of the sphere. Under the uplift of the spirit, the conscious functions leave go of the wheel of life and dip into the unconscious to join the fourth and form the One, the 'quintessence', the Diamond Body.

This represents to my mind a fresh and vivid image of the integration of the four functions. Not simply a lowering of the three towards the rising fourth, but rather an uplift of the spirit bringing all together.[6]

Viewed along the vertical axis, the resulting figure would appear uniform in colour, whole, a greenish gold. Viewed from the side, however, one would see only the luminosity of the gold. Manna personalities, charismatic figures, such as Christ and the Buddha, would appear so to 'the outside world'. Only along the divine or world axis would the true figure emerge, whether from above or below, as a blend of the gold of the sun with the blue and green of the earth. Perhaps only in such figures would all three conscious functions develop to that numinous gold.[7]

I realised that looking at the grey triangle as shadow would be an error. It is disproportionately small. The final inner triangle itself gave a better picture of the 'state of man': obscured areas of redeemed personal unconscious within a larger matrix of shadow. The divine possibilities, or opposites, seemed to be indicated by the formed gold triangles and the unformed, but suggested white triangles. But white, as a symbol for 'the devil' seemed very inapt. It was then I realised that Jung, in his *VII Sermones ad Mortuos*, had seen a rather similar thing. The devil is described as 'empty' or 'void', like the black hole I have described above as the opposite of the star, epitomised by the sun as symbolic of 'fullness'. White is not only completeness of colour, but also emptiness of colour, and the white triangles are not 'formed' and indistinguishable from the pleroma:

> Each star is a god, and each space that a star filleth is a devil. But the empty-fullness of the whole is the pleroma.   Sermo IV

Jung also associates the devil with the moon, the white 'hole' in the darkness of the night:

> The devil is the earth-world's lowest lord, the moon-spirit, satellite of the earth, smaller, colder, more dead than the earth.   Sermo 1 V

The circle, or the sphere, is the 'All', the pleroma, the macrocosm which is also our microcosm, since the small and the large are the same to the pleroma. Before the dawn of consciousness we stand in awe before it: the sun, the moon, the horizon, the cup. With the dawn we know that we are two, man and woman. We see its bounds and divide it into two. The division is straight, like our steps from the hearth to the door. The division gives us the measure of our domain, which we mark and claim. We perceive a centre. The centre is where all divisions 'cross'. It is a star. It is Holy. The star radiates; it pulses with life. But the centre is still.

The Star Mandala

The Birth

BEFORE I BEGAN painting what I came to call the 'diamond mandala' I had an urge to do some more doodling, completing in rough the lines which had seemed so full of significance when I was in the course of drawing the mandala. I was quite excited by one of the doodles which seemed to show what a sphere might look like full of tetrahedrons. So I decided to do some more precise work.[8]

The Perfecting

I confess I made use of modern technology in preserving the original drawing on a photocopying machine and so was able to continue where I had left off. When I had done this, the outer six-pointed star was revealed, with an inner hexagon surrounding my central triangle. It seemed necessary to draw in the complementary upright triangle, which then formed an inner star. I saw there was room for an outer hexagon, which seemed meaningful, connecting the points of the outer star, and drew that.

The Unfolding

I then perceived that, while certain lines might seem to suggest the full sphere, this was not visible in my drawing. On the other hand, I was excited by what I had drawn: a converging pattern of alternate hexagons and stars. I saw that I could continue the process *ad infinitum*. There seemed to be room for another inner pair, and so I drew them in.

I painted the resulting star mandala immediately after the diamond mandala, but did not immediately give it that name. I was still pursuing my goal of investigating the diamond as a three-dimensional Self symbol. I did not think much about the painting, but felt the outer rim portions to be ocean and the sky, and then moved naturally inwards rainbow-wise towards the red. I ran out of colours at the third hexagon. I then felt that the inner central hexagon must be left white, and it seemed natural to do the surrounding star and hexagon pale blue and violet. The result was a very pleasing mandala, but I did not make much of it. I inserted the birds and fishes to differentiate the sky from the ocean.[9]

The Creation of the Text

IMMEDIATELY AFTER painting the mandalas I had an urge to get down on paper the (to me) quite numinous series of inspirations that had accompanied the first mandala drawing, and the account above[10] is an edited version of that exercise. The following day I felt I had to continue by writing down the teeming ideas that were pouring into my head about the significance of the diamond. Strangely my first inspiration was to look up 'diamond' in the Oxford Book of Quotations, to see what I could find in literature, and I immediately hit on the nursery rhyme that heads this work. It was only then that it dawned on me that what I had done in the mandalas was to 'marry' the diamond and the star.

This experience continued to provide energy throughout my writing, which led me to research in the expected, but also some quite unexpected places, and I discovered surprise after surprise. What started as a short symbolic exercise turned into quite an extensive thesis.

The news which ignited the crisis my dreams were foreshadowing came in the midst of this work, eight days after I started writing. My writing became a creative outlet for all my feelings and emotion, and is very much a part of the whole experience. Dreams apart, the work was riddled with synchronistic experiences, which tend to come at times of emotional intensity.[11] Since the concept of synchronicity is so much a part of this account, I feel I should mention here some that I experienced outside the writing. Soon after the news reached me, and I was experiencing the pain quite intensely, there came to me the words of a psalm:

> who, walking in the vale of misery, use it for a well:
> and the pools are filled with water.   Psalm 84

Not long after, I decided to consult the *I Ching* on how I should cope. I received 48, 'The Well'! The third line was moving:

> The well is cleaned, but no one drinks from it.
> This is my heart's sorrow,
> For one might draw from it.
> If the king were clear minded,
> Good fortune might be enjoyed in common.

I felt a strange joy within me as I experienced this extraordinary 'coincidence'. At the time, I misread the new hexagram[12] as 8, 'Holding together'. I felt this to be meaningful and positive. It was only on checking my notes for the present purpose that I realised the error. The new hexagram should have been 29, 'Abysmal repeated'. If I had received this in the midst of my initial grief, I might have been despairing. It might have well signalled a depression. However, when I did get it right, it seemed doubly meaningful. The judgement reads:

If you are sincere, you have success in your heart,
And whatever you do succeeds.

And the image reads:

Water flows on uninterruptedly and reaches its goal.
Thus the superior man walks in lasting virtue
And carries on the business of teaching.

Again the image is of water, the water of life, quenching man's thirst. And the judgement could not be more encouraging at a point when the way ahead seemed without any clear goal. Yet I might well have closed the book without further study! The next day I went for a walk with a friend. As we set out, the bells were tolling for a funeral and I remarked, 'It's tolling for me!' When we came back later, my friend remarked that she felt that my depression was only a layer, and that underneath there was joy. As we arrived, the congregation were streaming out of church and I remarked, 'The funeral is over!'

So to me the mandalas represent the diamond and the star, essentially the six-pointed star inextricably entwined in embrace with the tetrahedrons of the Diamond Body of Chinese mysticism. In the second mandala each star is bounded by a hexagon, which means for me the wisdom of the hexagrams of the *I Ching*, the oldest book in the world. The pattern continues *ad infinitum*, representing the endlessly repeating cycles of life, the 'changes' of the great *Book of Changes*. The pattern circulates, moving to and fro from the macrocosm of the universe, the great All, through the centre of the body of man, the *tant'ien*, to the microcosm of the atomic nuclei at its foundation, where all is dancing energy, the Self.

*

Do not despair for Johnny head-in-air;
He sleeps as sound
As Johnny underground.

Better by far
For Johnny-the-bright-star,
To keep your head
And see his children fed.[12]

NOTES

1 'The Conundrum of the Workshops', Rudyard Kipling.

2 I later found a most significant reference to three keys in Cirlot: they symbolise a number of secret chambers filled with treasure and represent initiation and knowledge. The first key of silver concerns what can be revealed by psychological understanding; the second, of gold, represents philosophical wisdom; the third, of diamond, confers the power to act. Cirlot refers to 'eternal life', the key being to the gate of death. It was pointed out to me by Dr Richard Pope that life can be equated with 'the power to act'.

3 The Shri Yantra is often described as a mandala, and illustrated as an example in Cirlot's Dictionary of Symbols.

4 I was living in a small hotel in Küsnacht.

5 It might be thought that the apices are readily defined by the diameters of the circle. However only the horizontal diameter determines the positioning. The others were found by marking off radii from the horizontal tangents.

6 'The fourth [alchemical] nature ... leads straight to the Anthropos idea that stands for man's wholeness, that is the conception of a unitary being who existed before man and at the same time represents man's goal. The one joins the three as the fourth and thus produces the synthesis of the four in unity.' C.G. Jung, CW 12, 210.

7 Thinking over this, I believe I had in mind the rather special greenish tinge given to gold which has been flattened to an extreme thinness, almost to transparency. Greenish gold is mentioned in Jung, CW 12 (207) in a quotation from the alchemical text the *Rosarium: Our gold is not the commonest gold. Whatever is perfect in the bronze is that greenness only, because that greenness is straightway changed by our magistry into the most true gold.*

8 At the time and ever since I have been struck by the wish to create a three-dimensional solid structure representing these mandalas, but have been defeated. Perhaps the Chinese, skilled in the art of producing carved spheres within spheres, might manage it!

9 With hindsight it may appear that the results might have been achieved more easily. This, however, illustrates the difference between hindsight and creative work which never proceeds logically. I am well aware of this after working for so many years as a patent attorney. It often takes this critical eye to explain to an inventor what, in fact, he has created!

10 The original essay.

11 C.G. Jung, *Synchronicity*, Bollingen Series, 32, 33 (also CW 7).

12 Formed by converting the 'moving line' to its opposite.

13 'For Johnny', John Pudney.

Bibliography of Sources

Al-Huang, Chungliang, *Embrace Tiger, Return to Mountain*, Celestial Arts.
Al-Huang, Chungliang, *Quantum Soup*, Celestial Arts.
Arden, Margaret, *Midwifery of the Soul*, Free Association Books.
Campbell, Joseph, *Myths to Live By*, Viking.
Capra, Fritjof, *The Tao of Physics*, Flamingo.
Capra, Fritjof, *The Turning Point*, Flamingo.
Cirlot, *Dictionary of Symbols*, Routledge.
Clarke, J.J. *Jung and Eastern Thought*, Routledge.
Clarke, J.J. *Oriental Enlightenment*, Routledge.
Clarke, J.J. *The Tao of the West*, Routledge.
Conradi, Peter J., *Iris Murdoch: A Life*, HarperCollins.
Dawkins, Richard, collected quotations Richard Dawkins, interview.
Ellenberger, Henri F., *The Discovery of the Unconscious*, Harper Torchbooks.
Encyclopaedia Britannica.
Frey, C.T., Lectures at the Jung Institute, Zürich (unpublished).
Jung C.G., *Synchronicity*, Bollingen Series.
Jung, C.G., Collected Works, RKP, Vols 5, 6, 7, 8, 9 Part I, 12, 13, 14, 18
Jung, C.G., *Memories, Dreams, Reflections*, Collins and RKP.
Jung, C.G., *The Seminars*, Vol. 1 *Dream Analysis.*
Keating, Jacqueline, *Morning Star*, unpublished article submitted to the Jung
  Institute.
Lao Tsu, *Tao Te Ching*, various translations.
Larousse Encyclopedia of Mythology, Hamlyn.
Liddell and Scott, *Greek Lexicon*, Oxford, 1949.
Lovelock, James, *Gaia: The Practical Science of Planetary Medicine*, Gaia Books.
Lovelock, James, *Homage to Gaia*, Oxford.
Lovelock, James, *The Revenge of Gaia*, Penguin.
Mattoon, Mary Ann, *Jungian Psychology in Perspective*, Free Press.
Midgley, Mary, *Science and Poetry*, Routledge.
Midgley, Mary, *The Essential Mary Midgley*, Routledge.
More, Thomas, *Care of the Soul*, HarperCollins.
Murdoch, Iris, *The Sovereignty of the Good*, RKP.
Pelton, Robert, *Trickster in West Africa*, Berkeley Press, 1980.

Pirsig, Robert, *Zen and the Art of Motorcycle Maintenance*, Morrow.

Primack, Joel and Abrams, Nancy Ellen, *The View from the Centre of the Universe*, Fourth Estate.

Psalm 8, Revised Version.

Rose, Steven, *Lifelines*, Allen Lane.

Rose, Steven, *The Making of Memory*, Bantam.

Rozin, Poritsky and Scotsky, *Science*, March 26, 1971, 1264-7.

Schumacher, E.F. *Good Work*, Abacus.

St John's Gospel, 3, 8.

Tippett, Sir Michael, 'Feelings of Inner Experience', from *How Does It feel?*, Thames & Hudson.

Tracey, David, Lecture to the Scientific and Medical Network, 2006.

Ullman, M. [1973], 'Dreams, the Dreamer and Society', in *New Directions in Dream Interpretation*, State University of New York Press.

van der Post, Sir Laurens, *The Creative Pattern in Primitive Africa*, Eranos Lectures No 5.

van der Post, Sir Laurens, *The Lost World of the Kalahari*, Penguin.

Vickers, Salley, *The Other Side of You*, Fourth Estate.

Watts, Alan, *The Wisdom of Insecurity*, Pantheon.

Watts, Alan, *Tao: The Watercourse Way*, Penguin.

Watts, Alan, *What is Tao?*, New World Library.

Wilhelm, Richard, *I Ching or Book of Changes*, RKF.

Wilhelm, Richard, *The Secret of the Golden Flower*, RKP.

Wittgenstein, Ludwig, *Tractus Logica-Philsophica*, 4.026.

Index

Abrahamic religions, 6
adamantine, 42, 44, 45, 137
Adler, 98
Advent, 139, 146 (n.2)
Aids, 15
alchemists, 6, 9, 24, 26, 39, 117 (n.2),
 118, 121, 123, 126, 129 (n.4), 140,
 165, 172
alchemy, 6, 25, 62, 80, 84, 99, 104, 110,
 114, 115, 117 (n.2), 119, 127, 128,
 133, 148
Al-Huang, Chungliang, 13, 17 (n.1), 63,
 70 (n.25), 110, 117 (n.6), 142, 143,
 146, 157, 161 (n.13), 172
alimony, 45
'Alpha', 84, 86
alternative medicine, 15, 54
analogy, 50, 52, 55, 114, 146, 153
ankh, 173
Aphrodite, 84, 140
Arden, Margaret, 6, 12 (n.12), 82, 87-8,
 96 (n.1, 7-9), 97, 103, 105, 106,
 107 (n.1, 16), 158
Aristotle, 98, 108
astrology, 33 (n.11), 103-4
asymmetric thinking, 97
Attenborough, Sir David, 34
attitude
 cynical attitude, 49
 symbolic attitude, 10, 27, 30, 47-70,
  95, 100, 151, 156, 158, 159

Bateson, Gregory, 97
Baudelaire, 141, 146 (n.7, 8)
being, concept of, 19-20, 27-8, 30, 33
benzene 10, 134, 135
Bergson, 77
Binet, 103
Blanco, Matte, 97
Bohm, David, 106

Buddhism, 8, 9, 47, 85, 86
 see also Zen
Burghölzli Institute, 73, 75, 98
Bushmen, 126, 164-5
butterfly effect, 53-4, 155

Campbell, Joseph, 48, 81 (n.15), 88-9,
 96 (n.10)
Capra, Fritjof, xvi, 64, 65, 67, 70 (n.30),
 71, 119, 121, 129 (n.6-7)
carbon, xiv, xv, 46 (n.7), 57, 65, 80, 129,
 131-7, 172
 compound reactions, 135-6
 compounds, 131, 132, 134, 135-6
 crystalline phases, 57, 133
 energy diagram, 57, 134-5
 valencies, 129, 132
Cartesian partition – see Capra, Fritjof
caterpillar, 106-7
CAT scan, 57
central position
 of mankind in the universe, 90
chaos theory, 53
Chinese
 calligraphy, 62, 109, 123
 reading Chinese, 61-2, 167-8
Christianity, xiii, 4, 5, 15, 20, 23, 25, 26,
 32, 48, 58, 59, 84, 86, 114, 123, 127,
 139, 147, 152, 164
circle, 9, 48, 75, 108, 113, 114, 115, 120,
 124, 125, 127, 174, 176, 177,
 181 (n.5)
circulation around the centre, 114, 124-5
Clarke, J.J., 7-8, 47, 79, 81 (n.18), 110,
 117 (n.8), 124-5, 130 (n.17), 158,
 161 (n.14)
collective unconscious, xi, xiii, 22, 73,
 74-6, 78, 115
computer models, 57
Confucius, 63, 146

185

Ultimate Reality
Ken Moseley

'The author ... has amassed considerable scientific knowledge ... and then sat back to contemplate the pattern and meaning of reality with an informed, fresh and open mind'

Max Payne, Vice-President of the Scientific and Medical Network

Tapping the table at which they were sitting, the author asked his companions whether they had any idea of the ceaseless activity going on below the apparently solid surface. To his sceptical audience he explained that it was mainly empty space, the hardness being an illusion created by the rapid motion of the subatomic particles.

This led to a lively discussion about the nature of reality and our everyday perception of things. In an endeavour to clarify the issue, the author offered to explain in a letter, a letter which turned, four years later, into this book.

Thus began this voyage of discovery through the universe, amongst its atoms, solar systems and galaxies, to find answers to some of life's most basic questions. How does the universe work? What is our part in it? What is this intelligence we enjoy? Where does it come from and how are we able to ask, and even answer, such questions? Might this same 'knowingness' build the solar systems as well as the galaxies within which they evolve? Are there other universes and, if so, might they be part of an endless cycle of birth, maturity and death on a vast scale?

In jargon-free language, the author addresses these questions in the light of modern science, but by a process the author calls 'intuitive thought' he has expanded our understanding and broken down the barriers between different scientific disciplines to provide a 'theory of everything', which reconciles not only cosmology and particle physics but also the issue of consciousness, considered by many to be the most important unsolved mystery of modern science. He suggests that intuitive thought patterns are the language of cosmic reality and that the current limited worldview of our species can be expanded to levels that enjoy total knowingness, a oneness of mind and purpose with the Cosmos.

'Clearly you have written a work of vast scope'

Professor Lord Renfrew, Cambridge

ISBN 978 0 85683 237 6 · £14.95 · hb

For more information visit www.shepheard-walwyn.co.uk

MORE BOOKS FOR ENQUIRING MINDS

Marcus Aurelius: The Dialogues
Alan Stedall

'In this delightful and well-written book, Alan Stedall ... has done an enormous service in making some of Marcus Aurelius's reflections very accessible to the modern reader'   **Faith & Freedom**

'The Dialogues are eminently readable and immediate ... in places it is irresistible'   **The Philosopher**

Seeking an alternative to faith-based religion, Alan Stedall found in the *Meditations* of Marcus Aurelius rational answers to questions about the meaning and purpose of life that had been troubling him. Here too were answers to his concern that, in the absence of moral beliefs based on religion, we risk creating a world where relativism, the rejection of any sense of absolute right or wrong, prevails.

Inspired by the wisdom of the Roman Emperor, Stedall sought to present his reflections in a more contemporary and digestible way and employed the Greek philosophical technique of dialogue to create a fictional conversation between five historical figures who met in AD 168.

ISBN 978 0 85683 236 9 · £9.95 · hb

Wonders of Spiritual Unfoldment
John Butler

'This is an extraordinary book .... I have been reading it slowly since the moment when it became available, marvelling at the life experience, wisdom, and spiritual teaching.

'It seems to me that John expresses a very Christian spiritual vision in this book ... [yet] his openness to wisdom from any source, whether explicitly Christian or not, is surely a key to the understanding that he gains. Another striking aspect of this book is its lack of didacticism: John states at the beginning that he is not attempting to instruct his readers or to offer a spiritual self-help manual. Instead, he simply shares his rich experience and his endless sense of wonder at the mystery of God and his creation. I highly recommend this book for its fascinating, unusual, and deeply spiritual teaching. This is a book which, with huge generosity and honesty, offers these insights to all of its readers.'   **Mary Cunningham, Nottingham University, in *The Messenger***

ISBN 978 0 85683 260 4 · £15.95 · pb

For more information visit www.shepheard-walwyn.co.uk